How to Think with Numbers

How to Think with Numbers

Robert L. Hershey

William Kaufmann, Inc.
Los Altos, California

Library of Congress Cataloging-in-Publication Data

Hershey, Robert L., 1941-
How to think with numbers.

1. Mathematics—1961- I. Title.
QA39.2.H47 1982 510 81-17181
ISBN 0-86576-014-4 AACR2

ISBN 0-86576-014-4

10 9 8 7 6 5 4 3 2 1

Printed in the United States of America.

Contents

1. WHY SHOULD I THINK WITH NUMBERS?

The Story of Ug

Once upon a time there was a poor caveman named Ug. He was poor because he didn't know how to count past 5. You see, Ug's counting ability was limited to counting on the fingers of one hand. For any number bigger than 5 he ran out of fingers and he just said there were "many." Ug thought that his system was just great—easy and not too much thinking. There were times, though, that he had an inkling that not being able to count higher was costing him something. Even though poor Ug couldn't count past 5, he was an excellent fisherman. He was hardworking and patient and almost every day he caught 20 to 30 fish. Ug's neighbor, Slik, wasn't nearly as good a fisherman; he only caught six or seven fish per day. But Slik did know how to count!

This evening, as every evening after the day's fishing, Slik showed up at the entrance to Ug's cave holding six fish. He always timed his appearance to be there just as Ug came back from the lake with his big string of fish.

"Many for many," offered Slik. "Fair trade."

"Fair trade," said Ug, handing over 27 fish and accepting 6 fish in return.

Poor Ug made the same deal with Slik every day of his working life. He couldn't figure out why he and his wife, Nag, and his son, Ig, were always feeling hungry. After all, they had "many" fish to eat, just like Slik's family, who always seemed to be fat and contented. Poor Ug suspected that the problem might have something to do with counting past "many." He never was able to figure it out until his dying day, which came quite a bit earlier than it should have because of malnutrition.

Winners and Losers

It's been several thousand years since the time of Ug but the same kinds of things are still happening today. People are still being ripped off because they don't know how to think with numbers. These people are constantly being gypped, losing money, and missing opportunities. They are always falling for bad deals like installment purchases, gambling, and life insurance. We'll discuss the hows and whys later in this book.

On the other hand, there are many people who can think with numbers and are quite good at it. These people generally get ahead easily, have good jobs, make good investments, and have lots of money. They are usually well prepared for whatever happens. If an opportunity comes along they know how to use it, and if they get a bad break they come out of it smelling like a rose.

It's easy to tell the winners in life from the losers. Who wins and who loses is not just a question of basic intelligence or a question of luck. It depends to a great extent on how well you can think with numbers. If you're already a winner, this book will show you how to win more by thinking more effectively with numbers. If you've been losing lately, this book will show you why and get you thinking with numbers like a winner.

Pocket Calculators

Within the last few years there has been a revolutionary development that has made it much easier to think with numbers—inexpensive pocket calculators. Electronic pocket calculators have become much cheaper to make because of advances in "integrated circuit" manufacture, which put a whole circuit on a tiny chip. At the time of this writing, calculators that will add, subtract, multiply, and divide are being sold for $10 to $20. At this price everyone should be able to buy a calculator; you can't afford not to. The money you save by being able to think with numbers should easily return your investment within a few months.

A few years ago people might have made excuses that they couldn't think with numbers because pencil-and-paper arithmetic took too long or because they weren't very good at computation. These excuses no longer hold water. Today you don't have to be good at arithmetic to think with numbers. The calculator does the computation for you in a fraction of a second. What you do have to be good at is setting up the calculations so that all that's left is arithmetic computations for the calculator. This book will help you learn how to set up calculations.

How to Buy a Calculator

Electronic pocket calculators are a real bargain today. Even though everything else seems to be going up in price in this age of inflation, calculators have actually gone down in price over the past several years. When they first came out, even the cheapest ones cost well over $100. Because of the immediate popularity of calculators, there was much competition among electronics manufacturers, and better, less expensive methods were developed to make them. Nowadays you should be able to get a calculator for $10 to $20 depending on the features you want.

One of the most useful types of calculators is the credit-card-size calculator, which is convenient to carry with you in pocket or purse. Generally these come with a long-life battery that will give about 1000 hours of continuous operation.

Any calculator you buy should have the capability to add, subtract, multiply, divide, and place the decimal point. Any features beyond that will cost more. You will not need the added features for any of the problems in this book or for most other applications. But if you can afford to buy a more expensive calculator and you think you might use the features, then go ahead and buy it. It's a good investment.

The first thing you can spend money for is more accuracy. This means more digits in the display. The problems in this book were done on a calculator with an 11-digit display, but a display of at least 8 digits will be good enough for most practical purposes. If you can afford a calculator with more digits, get it. Anyway, you should be able to round off the answers in this book to match the answers you get on your own calculator, no matter what accuracy it has.

Some of the more expensive calculators also have provisions for expressing numbers that are very small or very large by using "scientific notation." Scientific notation is nice to have on a calculator but it's not a necessity. We will talk about scientific notation in Chapter 4.

Often calculators have provisions for storing numbers. This makes some problems easier to do since you don't have to write down intermediate numbers on a piece of paper, but can instead store them in the memory of the calculator. For the problems in this book, we won't expect you to have a storage feature on your calculator.

Our discussion has covered just the basic features you can buy in a calculator. There are other much more expensive "scientific" calculators, which have many more features. Unless you are an engineer or a scientist, the extra features in these calculators probably won't be worth the additional expense.

In buying your calculator your best bet is to compare several and find the one that has the features you want at the lowest price. Generally you can find a display of several calculators at department stores, business machine dealers, electronics shops, and bookstores. Try out the various calculators at the counter, if they let you, and look at the instruction booklets. Also ask your friends who have calculators how they like their particular model. Then compare prices and buy one.

If you already own a calculator, then we don't have to tell you how useful they are. You're all equipped to get the full advantage of this book.

How to Get the Most Out of This Book

This book will show you how to think with numbers, which is also called analysis. Analysis consists of looking at a problem stated in words and figuring out what computation is necessary to find the answer. In the old days when you were in grammar school you probably called it "word problems." Perhaps you were very good at word problems. Or perhaps you didn't understand them very well because you had a poor teacher, or because you hadn't yet had enough experience making and spending money to visualize their application, or because you got lost in the arithmetic of the computations. If you didn't understand then, you have a much better chance now. This book will be your teacher. From the explanations of the sample problems you will immediately find applications to your daily life. If you work out these problems using your pocket calculator, arithmetic should be no problem.

Here's how to get the most out of this book.

1. **Buy a Pocket Calculator**
   At a minimum you will need an electronic calculator that can add, subtract, multiply, and divide.

The use of a calculator is essential to getting the most out of this book. If you buy your first calculator now and use it to learn how to do the problems in this book, you will find it one of the best investments of your life.

2. **Learn How to Work Your Calculator**
   Read the literature that comes with your calculator and play with some calculations until you are sure you know how to add, subtract, multiply, and divide. We will assume that you know how to do these arithmetic operations with your calculator when we go through sample problems.
3. **Start from the Beginning of This Book and Go All the Way Through**
   As you go through the book, work out each sample problem, making sure you understand how the problem was set up. Do the arithmetic on your calculator and make sure you get the same answer as the book. Always have a pencil and paper handy for calculations. You may also want to make marginal notes in this book to remember your thoughts on the various sample problems. Try every problem yourself first before reading the solution in the book. Stick with every problem until you thoroughly understand it before going on to the next one, rereading as necessary. If the problems in the early chapters seem too easy you can feel flattered that you're smarter than the average reader. Stay with it—the problems will get more challenging as you go along. Probably as you go through there will be new calculations you haven't seen or new approaches that will simplify something you wondered about.

How This Book Will Help You

This book will show you how to analyze quantitative problems that come up again and again in your daily life. These are problems that you need to know how to solve for your economic survival. Sure, if someone else has done the analysis you might be tempted to take his word for it and avoid the labor of thinking. But then Ug took Slik's word for it too, didn't he?

Here is specifically what you're going to learn in this book.

1. **Percentages.** You will never again be fooled by percentages stated in a misleading way.
2. **Probabilities.** This is what you have to understand so that you don't get ripped off in gambling games or in buying insurance you don't need.
3. **Big Numbers and Small.** This chapter will review the naming of very large and very small numbers. You will also be introduced to scientific notation so that you can easily work with very large and very small numbers.
4. **Money and Time.** The important concepts of first cost and operating cost are discussed. You'll need these for your budgeting.
5. **Interest.** You will learn to calculate the cost of mortgages and installment payments and see how the bank figures interest on your money.
6. **Trade-Offs.** In many situations you have to give up some of something to get more of something else. This chapter shows you how to find out what you're getting.

Once you have gone through this book and worked the sample problems you will find yourself thinking with numbers and getting ahead. Anyone who can't think with numbers in an age when everyone has a calculator will soon find himself left behind.

2. PLAY THE PERCENTAGES

Markups

There's an old story about a man who ran a successful women's clothing store for 20 years despite the fact he had never finished grammar school. He was determined, however, that his son would have all the advantages, so he sent him off to college in Boston and then to Harvard Business School. Armed with his shiny new M.B.A., the son returned and wanted to help in the family business.

"How much is your markup percentage?" asked the son.

"Five percent," said the father.

"Holy cow!" said the son. "How can you make any money with such a ridiculously low markup percentage?"

"Well, it always seems to work," said the father. "I buy a bra for $1 and sell it for $5. That's 5%!"

The joke, of course, is that the father knew nothing about percentages. He thought 5% meant five times. He had been setting prices on the basis of his experience of what the public was willing to pay. This ancient story obviously came before the time of income taxes. It is inconceivable today that any businessman could remain in business without understanding percentages.

The markup percentage in the story was 400%, not 5%. The father added $4 to the wholesale price of $1, which he had paid for the bra. Since $4.00 ÷ $1.00 = 4.00, the markup was 400%.

The old rule for finding percentages that you learned in grammar school went something like this.

To find the percentage express the number as a decimal fraction. Read everything to the left of the decimal point and the first two places to the right of the decimal point and call it percent.

The rule is sometimes stated as, "Move the decimal point two places to the right and call it percent." Use whichever way is easier for you to remember. The result is the same.

In the story the decimal fraction turned out to be 4.00. Reading the number 4 to the left of the decimal point and the two 0s to the right, we get 400, and we call it percent.

You might also remember that percent means hundredths. So 1% is the same as 1/100. Similarly, 10% is 10/100 or 1/10. Another percentage often used is 25%, which is 25/100 or 1/4. Similarly, 50% is the same as 1/2.

Suppose the story were changed a little, and the father had said that he bought the bra for $1.37 and sold it for $6.85. Is the percentage the same? Take out your calculator and find out.

In working this problem and all the others in this book, look at the problem first to figure out what information is given and what number must be found. Then set up the calculation, do the arithmetic on your calculator, and compare your answer with the one in the book. If you don't get the same answer, then study the problem to see what you did wrong.

Problem 1

A man buys a bra for $1.37 and sells it for $6.85. What is his markup percentage?

Given:

Wholesale price = $1.37
Retail price = $6.85

Find:

Markup percentage.

Solution:

The amount of the markup is the retail price minus the wholesale price.

$$\$6.85 - \$1.37 = \$5.48$$

To find the markup as a decimal fraction, we divide the amount of markup by the wholesale price.

$$\$5.48 \div \$1.37 = 4.00$$

This division should be very easy with your calculator. The decimal fraction 4.00 is, of course, equivalent to 400%.

Answer:

Markup percentage = 400%.

Thus we see that the new numbers happen to give the same answer, 400%. The original numbers $1 and $5 make a better joke, though, since not everyone will have a calculator handy.

Okay, let's try another markup problem.

Problem 2

A used-car dealer buys a car at an auction for $322.25. Later he sells the car to a customer for $598.50. What is the markup percentage?

Given:

Wholesale price = $322.25
Retail price = $598.50

Find:

Markup percentage.

Solution:

The form of this problem is just like Problem 1, but the numbers have been changed. Thus the set up of the problem is exactly the same, but the numbers you punch on your calculator are different.

Again the amount of the markup is the retail price minus the wholesale price.

$$\$598.50 - \$322.25 = \$276.25$$

The markup as a decimal fraction is found from

$$\$276.25 \div \$322.25 = 0.857253685$$

If your calculator doesn't give this many decimal places don't worry about it. For all practical purposes, we round off to 0.857 anyway.

Answer;

Markup percentage = 85.7%

Notice that in Problem 2 the 7 in the 85.7% was three places to the right of the decimal point when the markup percentage was expressed as a decimal fraction. Therefore, in the final answer the 7 was tenths of a percent.

Notice also in Problem 2 that the numbers were rather long, and not nice round numbers. In real life this is the way numbers usually are. If this problem had been done with pencil-and-paper arithmetic rather than with a calculator, it probably would have taken at least 10 minutes to do the "long division." With a calculator, though, the problem was as easy as the first one. This is the beauty of calculators. Now you can think with numbers without the hassle of a lot of time-consuming paperwork.

Let's take another look at the used-car situation of Problem 2, but this time from the customer's point of view.

Problem 3

Sam Smith sees a used car on a dealer's lot and buys it for $598.50. What percentage of his purchase price could he have saved if he had bought it instead at the auction for $322.25?

Given:

Wholesale price = $322.25
Retail price = $598.50

Find:

Percentage of retail price saved by buying wholesale.

Solution:

The amount of money that Sam Smith could have saved by buying at the auction is

$$\$598.50 - \$322.25 = \$276.25$$

The fraction of the purchase price he could have saved is

$$\$276.25 \div \$598.50 = 0.4615705931$$

This rounds off to 0.462.

Answer:

Percentage of retail price he could have saved = 46.2%.

Arguing with Percentages

In Problems 2 and 3 we were working with the same price difference but we came up with two different percentages. The difference was the number we used for the base. The base is the number we divide by when we take the decimal fraction. It's the number that answers the question, "Percentage of what?" How do we choose which number to use as the base? In these problems the choice was easy because it was contained in the statement of the problem. In real life the choice is more difficult. We choose the base that seems best for a meaningful comparison. In Problem 2 the auction price was the meaningful base for the used-car dealer in figuring his markup. In Problem 3 the price Sam Smith paid is the more meaningful base from the customer's point of view. The percentage calculated with the smaller base of $322.25 in Problem 2 was 85.7%. The percentage calculated with the larger base of $598.50 in Problem 3 was 46.2%. Thus if the amount we are dividing into is the same in each case, the percentage comes out larger for the smaller base.

The fact that percentages always come out larger when referred to a smaller base frequently leads to misleading statements about percentages by those who have an ax to grind. The next three problems will illustrate this process in considering the case of a candy store owner who has raised the price of a candy bar from $0.15 to $0.20.

Problem 4

The price of a candy bar that is now $0.15 goes up by $0.05. What is the percentage of increase in price?

Given:

Price = $0.15
Increase = $0.05

Find:

Percentage of increase in price.

Solution:

The percentage of increase in price is

$$\$0.05 \div \$0.15 = 0.3333333333$$

This rounds to 0.333.

Answer:

Percentage of increase = 33.3%.

This percentage of increase seems like a pretty high number compared to the original price. The candy store owner may, however, argue to John Robinson, who buys a candy bar per week, that this change is just a few pennies and that the way to see how it will affect John's life is to figure the price increase as a percentage of John's weekly salary.

Problem 5

John Robinson buys a candy bar every week. The price of the candy bar, which was previously \$0.15, increases by \$0.05. John Robinson's salary is \$223.50 per week. What percentage of John's weekly salary will be required to pay for the candy bar price increase?

Given:

Weekly salary = \$223.50
Candy bar price increase = \$0.05
Candy bar original price = \$0.15

Find:

The price increase as a percentage of the weekly salary.

Solution:

The required decimal fraction is

$$\$0.05 \div \$223.50 = 0.0002237136465$$

We round this to 0.000224.

Note that the original price of the candy bar was extra information. We didn't need it at all for this problem. That's the way real problems often are. You have to figure out what information to throw away.

Answer:

The candy bar price increase as a percentage of John Robinson's weekly salary is = 0.0224%.

This is a very small percentage of John Robinson's salary. But how about his six-year-old son, Junior? Junior's allowance is \$0.25 per week. How does the price increase affect him?

Problem 6

Junior receives an allowance of \$0.25 per week. He buys a \$0.15 candy bar every week, whose cost is now increasing by \$0.05. What percentage of his weekly allowance must go to absorb the increase?

Given:

Junior's weekly allowance = \$0.25
Candy bar price increase = \$0.05
Candy bar original price = \$0.15

Find:

The candy bar price increase as a percentage of Junior's weekly allowance.

Solution:

The decimal fraction required is

$$\$0.05 \div \$0.25 = 0.2$$

Again, the original price of the candy bar was just extra information.

Answer:

The percentage of Junior's allowance that must go for the price increase = 20%.

Thus in looking at various bases we found that the candy bar price increase was 33.3% of the original price, 0.0224% of John Robinson's salary, and 20% of Junior's weekly allowance. It is hard to say which percentage is the most meaningful, but most people would say it's the increase over the original price.

Finding the Amount from the Percentage

All the problems so far have been those where we find the percentage. There are also many problems where we are given a percentage and must find an amount, as in the next problem.

Problem 7

Henry Jones earns \$256.40 per week. One week he earns a bonus of 32.8% of his weekly salary. How much money is the bonus?

Given:

Henry Jones's weekly salary = \$256.40
Bonus = 32.8% of weekly salary

Find:
Amount of bonus.

Solution:
First convert the percentage to the decimal fraction 0.328.

To find the amount of the bonus, multiply the decimal fraction by the weekly salary.

$$0.328 \times \$256.40 = \$84.0992$$

Rounding it to the nearest whole cent gives $84.10.

Answer:
The amount of Henry Jones's bonus = $84.10.

The same type of calculation comes up again when we calculate the taxes on Henry Jones's bonus.

Problem 8

Henry Jones is in the 27% income tax bracket. He earns an additional $84.10 this year. How much of the money will the government get?

Given:
Tax percentage = 27%
Additional earnings = $84.10

Find:
Amount that is paid in taxes.

Solution:
First convert the percentage to the decimal fraction 0.27.
Then multiply by the amount of additional earnings.

$$0.27 \times \$84.10 = \$22.707$$

Rounding to the nearest whole cent gives $22.71.

Answer:
Henry Jones's taxes on the additional earnings = $22.71.

Percentage of Growth

Up to now we have talked about percentages used in a single calculation. When we are calculating percentage of growth, the percentage calculation must be repeated several times. This type of problem comes up often in finding how much the value of your house will grow in several years or how much your salary will increase over several years.

To see how this process works, let's try a problem.

Problem 9

Brian Van Buren owns a house whose value is growing by 9% per year. It is now worth $60,000. How much will it be worth in 1 year? In 2 years?

Given:

Original value of house	= $60,000
Percentage of growth	= 9%
Number of years	= 2

Find:

Amount the value grows over the first year.
Total value of the house in 1 year.
Amount the value grows over the second year.
Total value of the house in 2 years.

Solution:

Over the next year the amount the value of the house grows is simply the percentage of growth times the present value of the house.

$$0.09 \times \$60{,}000 = \$5400$$

If we add this to the original value of the house, we obtain the total value of the house in 1 year.

$$\$60{,}000 + \$5400 = \$65{,}400$$

Now let's repeat the process by finding the amount the house grows over the second year. To find it we'll multiply the value at the end of the first year by the percentage of growth.

$$0.09 \times \$65{,}400 = \$5886$$

Then the total value at the end of the second year is the value at the end of the first year plus the growth for the second year.

$$\$65,400 + \$5886 = \$71,286$$

Answer:

Amount the value grows over the first year = $5400.

Total value of the house in 1 year = $65,400.

Amount the value grows over the second year = $5886.

Total value of the house in 2 years = $71,286.

Now that we've worked the problem, let's look at an easier way we could have used if all we were trying to find was the total value after 2 years. The rule presented below can be used in problems involving percentage of growth.

To find the total value at the end of a number of years (n), multiply the original value by (1 + percentage of growth) n times.

Let's test this rule on the problem we just solved and check that it gives the same answer.

In this problem (1 + percentage of growth) was (1 + 0.09). We have indicated that the sum of 1 and the percentage of growth are grouped together as a single quantity by writing them in parentheses. If we add them, we have

$$(1 + 0.09) = 1.09$$

To follow the rule stated above, we multiply the original value by 1.09 twice, since we want the value at the end of 2 years.

$$\$60,000 \times 1.09 \times 1.09 = \$71,286$$

Multiply this out on your calculator and check that it works.

Now that we know this rule, let's try it on another problem about another house with a lower growth rate.

Problem 10

Fred Thomas owns a house which grows in value by 4% per year. It is now worth $42,600. How much will it be worth in 4 years?

Given:

Present value of house = $42,600
Percentage of growth = 4%
Number of years = 4

Find:

Value of house in 4 years.

Solution:

Since the growth is over 4 years we have to multiply the present value by (1 + percentage of growth) four times.

$$\$42{,}600 \times (1 + 0.04) \times (1 + 0.04) \times (1 + 0.04) \times (1 + 0.04) = \$49{,}835.97466$$

Rounding the answer to the nearest cent gives $49,835.97.

Answer:

In 4 years Fred Thomas's house will be worth $49,835.97.

The process isn't much different for a different number of years. We just keep multiplying by (1 + percentage of growth) as many times as there are years. This is pretty easy with a calculator.

You may have once learned that the way to show that a number is multiplied by itself a certain number of times is with a little number, which is higher up than the other numbers. Thus in the previous problem we could have saved some writing by saying

$$(1 + 0.04) \times (1 + 0.04) \times (1 + 0.04) \times (1 + 0.04) = (1 + 0.04)^4$$

It means exactly the same thing as writing out (1 + 0.04) four times with multiplication signs in between. The little number above the line is called an exponent. The number $(1 + 0.04)^4$ is read as "the quantity one plus zero point zero four to the fourth power."

Let's try another problem with percentage of growth which will have a larger number of years.

Problem 11

Charles Warren is doing very well on his job and he expects to get a raise of 7% every year. If he makes $8000 per year today, how much can he expect to be making in 10 years?

Given:

Present salary = $8000
Percentage of growth = 7%
Number of years = 10

Find:

Amount of salary in 10 years.

Solution:

We can use the notation of the exponent to show that (1 + percentage of growth) is multiplied by itself 10 times.

$$\$8000 \times (1 + 0.07)^{10} = \$15{,}737.21086$$

Rounding to the nearest whole cent gives $15,737.21.

Answer:

Charles Warren's expected salary in 10 years = $15,737.21.

Notice that this last problem involved some work since we had to punch 1.07 on the calculator 10 times. But think how long it would have taken without a calculator. Later in Chapter 6 we will learn how to do problems like this somewhat faster using the tables in the back of this book.

The Rule of 70

In Problem 11 you will notice that Charles Warren's predicted salary in 10 years came out about twice what he is making now. Often we are interested in the period of time it takes a quantity to double using a certain percentage of growth. There is a good rule of thumb which can be used to find the doubling period very quickly. It is called the Rule of 70 and it is stated as follows.

To find the doubling period divide the percentage of growth into 70.

For example, in Problem 11, the percentage of growth is 7%. To apply the Rule of 70 we do *not* convert the percentage of growth into a decimal fraction (this is the only time we don't). Instead we divide the number 7 itself into 70.

$$\text{Doubling period} = 70 \div 7 = 10 \text{ years}$$

We now see why Charles Warren's salary in Problem 11 happened to come out about twice his present salary. It was because 10 years was indeed the doubling period.

Let's now apply the Rule of 70 to some problems.

Problem 12

Tom Williams puts $500 in a savings and loan association that pays 6% interest. When can he expect it to have doubled?

Given:

Percentage of growth = 6%
Present amount = $500

Find:

Doubling period.

Solution:

The fact that he put in $500 was just extra information. To find the doubling period use the Rule of 70.

$$\text{Doubling period} = 70 \div 6 = 11.66666667 \text{ years}$$

Round it to the nearest tenth of a year and call it 11.7 years.

Answer:

Doubling period = 11.7 years.

The Rule of 70 can also be used to find the percentage of growth necessary to double an amount in a certain number of years. To do this we divide the doubling period into 70 and get the percentage of growth. This will be illustrated in the next problem.

Problem 13

Bill Brown bought his house 8 years ago for $20,000. It is now worth $40,000. What was the annual percentage of growth in the value of his house?

Given:

Previous value = $20,000
Present value = $40,000
Period = 8 years

Find:

Annual percentage of growth.

Solution:

Since $40,000 ÷ $20,000 = 2, the house has doubled in value. If the house had not doubled but had some other relation, we would not be able to use the Rule of 70. We would have to use methods to be discussed later in Chapter 6.

Using the Rule of 70 gives

$$\text{Percentage of growth} = 70 \div 8 = 8.75$$

Answer:

The percentage of growth is 8.75%.

In this chapter we have reviewed percentages and become familiar with setting up the problems and working with the calculator. We're sure many of you will have found this chapter too easy. However, percentages are used quite often in this book and we wanted to make sure everyone knows them before going on to the next chapter.

3. WHAT ARE THE ODDS?

Heads and Tails

Suppose you are playing heads and tails for money. If the coin comes up heads you win $1; if it comes up tails you lose $1. That seems like a fair game, doesn't it?

Now, let's change the game a little. If it comes up heads you win $1. If it comes up tails you lose $2. You say that's not a fair game? You're right! Other gambling games are rip-offs too. In this chapter you'll learn to calculate the odds and you'll find out exactly how much you're getting ripped off.

When we are talking about odds we are really talking about probability. *The probability is the percentage of the time that an event happens in the long run.*

In our coin-flipping example the probability of the coin coming up heads is 50%. This means that if we flipped the coin a large number of times we would find that roughly half the flips had resulted in heads. For instance, if we flipped the coin 2000 times we would get heads around 1000 times and tails around 1000 times. For a very large number of flips the percentage of heads will be very close to 50%. The probability can also be expressed as the decimal fraction 0.5.

Mathematical Expectation

Related to the idea of probability is the problem of figuring out how much your chances are worth. The value of your chances in a game is called the "mathematical expectation." We're using the term game very broadly to mean anything where chance and decisions are involved, such as buying insurance or making career choices. We're not just talking about things like coin flipping and dice.

The mathematical expectation of gain is equal to the probability of winning times the payoff for winning minus the probability of losing times the amount to be lost.

For example, in the coin-flipping game that you knew wasn't fair, the probability of winning was 0.5 and the amount you would get for winning was $1. The probability of losing was 0.5 and the amount you would lose was $2. These are the only two possibilities. The mathematical expectation of gain is

$$(0.5 \times \$1.00) - (0.5 \times \$2.00) = -\$0.50$$

Notice that the mathematical expectation of gain is negative. This means that this is a bad bet and you will lose money in the long run. The fact that the mathematical expectation of gain is −$0.50 means that in this game you will lose an average of $0.50 every time you play.

The original game, which you thought was fair, had equal payments for winning or losing of $1 each.

Thus the mathematical expectation of gain for that game was

$$(0.5 \times \$1.00) - (0.5 \times \$1.00) = 0$$

Since the mathematical expectation of gain is zero, this is indeed a fair game. In the long run the average amount you win or lose per game becomes small.

From the concept of mathematical expectation, it is easy to see how to win in the long run. Only bet when the mathematical expectation of gain is positive. We will see later in this chapter that all commercial gambling games have negative mathematical expectations of gain. This includes casino gambling with dice, roulette, slot machines, and cards. It also includes pari-mutuel betting on horse races, dog races, and athletic events. In these games you have to lose in the long run, so don't play.

Just as there are commercial gambling rip-offs that take advantage of people who are too eager to take chances, there are also insurance rip-offs that take advantage of people who are afraid to take chances. Let's try a problem and see if you can figure out the mathematical expectation of gain.

Problem 14

Roger Young has just bought a new car for $3962. He is considering whether to buy a $100-deductible collision insurance policy that has an annual premium of $245. He expects that during the course of the year the probability of "totaling" his car in an accident is 0.01. He also expects that the only other type of accident might be a "fender-bender" that would do $500 worth of damage with a probability of 0.25 during the year. Roger has lots of money in the bank and he could buy another car immediately if this one was totaled. What is Roger's mathematical expectation of loss if he doesn't buy the collision insurance? What is it if he buys the insurance?

Given:

Value of car	= $3962
Probability of total loss of car	= 0.01
Damage from a "fender-bender"	= $500
Probability of a "fender-bender"	= 0.25
Deductible level	= $100
Premium	= $245

Find:

Mathematical expectation of loss in not buying collision insurance.

Mathematical expectation of loss in buying collision insurance.

Solution:

In this "game" we are calculating the mathematical expectation of loss instead of gain because there is no way Roger can come out ahead. He can lose his car through accident or his money through insurance premiums, but there is no way he can come out with more cars or more money. We want to see which choice loses him the least.

The calculation procedure is the same in figuring a mathematical expectation of loss as it was for gain; we just avoid the necessity of putting negative signs on all the dollars.

First let's look at the case where Roger doesn't buy insurance. There are just three things that can happen by his picture of the situation.

(1) Total loss accident: probability	= 0.01
(2) "Fender-bender" accident: probability	= 0.25
(3) No accident: probability	= 0.74

Note that the probability of no accident has to be 0.74 in this problem, because the probabilities of all the possible outcomes must add up to 1.

$$0.01 + 0.25 + 0.74 = 1.00$$

According to the statement, totaling loses Roger $3962, "fender-bending" loses him $500 and, of course, there is zero loss when there is no accident. Multiplying the loss associated with each outcome by its probability and adding gives the mathematical expectation of loss with no insurance.

$$\$3962 \times 0.01 + \$500 \times 0.25 + 0 \times 0.74 = \$164.62$$

Next, let's look at his mathematical expectation of loss in buying insurance. He has the same three possible outcomes with the same probabilities as with no insurance—totaling, "fender-bending," or no accident. The difference this time is that the accidents will cost him only the $100 deductible. He also will lose his $245 premium no matter what happens; this has a probability of 1.0. Adding these up we get the total mathematical expectation of loss when Roger buys insurance.

$$\$100 \times 0.01 + \$100 \times 0.25 + 0 \times 0.74 + \$245 \times 1.0 = \$271$$

Answer:

Roger's mathematical expectation of loss is $164.62 if he doesn't buy insurance.

It is $271 if he does buy insurance.

Thus it turns out that Roger would be better off not buying the insurance. In this case, Roger had money in the bank so he was able to insure himself against the loss of his car at a more favorable rate than an insurance company would give him since their rates also included handling costs and profit.

The above example was very much simplified in that we only considered two kinds of accidents. The results, though, were typical of real life. If you have enough money to insure yourself, don't buy collision insurance.

Let's look at another example.

Problem 15

Harriet Powell likes to gamble on "the numbers." Every day she picks one number out of a list of 1000 possible; the numbers go from 000 to 999. Since all the numbers are equally likely, the probability of her picking the winning one is 1 in 1000, which is 0.001.

She bets $2. If she wins she gets $1000. What is her mathematical expectation?

Given:

Probability of winning = 0.001
Amount of payoff for winning = $1000
Amount of bet = $2

Find:

Her mathematical expectation.

Solution:

The probability of Harriet losing the original $2 is 1.0 since she does not get that back even if she wins. The mathematical expectation is again equal to the probability of winning times the amount she wins minus the probability of losing times the amount she loses.

$$(0.001 \times \$1000) - (1.0 \times \$2.00) = -\$1.00$$

Answer:

Her mathematical expectation is −$1.00. On the average, she loses $1 every day she bets $2.

The above example of a commercial gambling game is typical of the negative mathematical expectations involved for the player. The mathematical expectations for the house are, of course, positive. This is the reason why many states run lotteries as revenue-producing measures.

Let's look at another gambling example, this time with horse races.

Problem 16

Bradley Rice is betting on a match race between two horses, Submarine and Whirlagig. Both horses have odds of 4 to 5. Bradley puts $2 on Submarine. What is his mathematical expectation? (Assume that the probability of either horse winning is 0.5.)

Given:

Amount of bet = $2
Payoff odds = 4 to 5
Probability of winning = 0.5

Find:

Mathematical expectation.

Solution:

The payoff odds of 4 to 5 mean that Bradley will win four-fifths of $1 or $0.80 for every $1 he puts up when he picks the winning horse. Thus if Submarine wins he will win $1.60 and also get his $2 back. If, instead, Whirlagig wins, which has a probability of 0.5, he will lose $2.

The mathematical expectation is the probability that Submarine wins times the payoff if Submarine wins minus the probability that Whirlagig wins times the loss if Whirlagig wins.

$$(0.5 \times \$1.60) - (0.5 \times \$2.00) = -\$0.20$$

Answer:

His mathematical expectation is −$0.20.

Most horse races are not match races like the above example, where there are only two horses running. This example was chosen to simplify the calculation and illustrate the negative mathematical expectation of gain. The mathematical expectation of gain is negative because the operators of the racetrack always take out a "mutuel take" of about 20% off the top before calculating the odds. This pays for their own expenses and profits and for the state's taxes. Thus the bettors are in effect betting against the other bettors with some of the money missing. The only chance a bettor has for a positive mathematical expectation of gain is if he somehow has better information than the other bettors. Such "special knowledge" is something you find in tall tales, not reality. But millions of horseplayers continue to lose, year after year, mistakenly thinking they are smart enough to beat the system when they are reading the same racing form as everyone else.

Besides figuring out betting situations, mathematical expectation can be used in other situations where decisions must be made. Often people are called upon to make career decisions that involve taking some chances. Success in the new job may depend on some events that are unsure, like making sales. Let's try a problem on this.

Problem 17

Walter Crawford is thinking of changing jobs. He is presently a clerk making $8335 per year. If he becomes a salesman he will be on 10% commission rather than on salary. The most he could expect to sell is $200,000 worth of the product, but he honestly estimates there is only a 10% chance of his selling this much. He thinks that there is a 20% chance that he could sell $150,000 worth and a 50% chance that he could sell $100,000 worth. There is also a 20% chance that he will do very badly and only sell $50,000 worth. What is his mathematical expectation of his annual commission in the new job? How does that compare with his present salary?

Given:

Present salary = $8335
Commission rate = 10%
Probability of selling $200,000 worth = 0.1
Probability of selling $150,000 worth = 0.2
Probability of selling $100,000 worth = 0.5
Probability of selling $50,000 worth = 0.2
(Note that the probabilities add up to 1.0.)

Find:

Mathematical expectation of Walter Crawford's annual commission.

Solution:

Since the commissions are 10%, Walter gets $20,000 on $200,000, $15,000 on $150,000, $10,000 on $100,000, and $5,000 on $50,000.

The mathematical expectation of his commission is thus

$$(0.1 \times \$20{,}000) + (0.2 \times \$15{,}000) + (0.5 \times \$10{,}000) + (0.2 \times \$5{,}000) = \$11{,}000$$

In comparing this to his annual salary of $8335, we see that the new job with the commission is likely to give him more income.

Answer:

The mathematical expectation of the gain in income is $11,000 − $8335 = $2665.

In Problem 17 Walter found that on the average he would make $2665 more per year, and he decided to change jobs. The principle is simple—*do things that give you positive mathematical expectations of gain and avoid things that give you negative mathematical expectations of gain.*

The key to making good decisions based on mathematical expectations is to be honest with yourself in estimating the probabilities. Find out all the facts you can and come up with the best numbers that you can for the probabilities. If you use mathematical expectation to make decisions like this involving large sums of money, you are, of course, taking a chance. You may have bad luck on your one try and be one of the people who lose in those circumstances. If you think you can't live with that big a loss, don't take the chance. On the other hand, you may have good luck and win more than the mathematical expectation of gain would indicate. People, on the average, will come out as the mathematical expectation says.

Probabilities of Several Events All Happening

One thing that sometimes helps in determining probabilities is to know about the probability of several events all happening. One example of this is the roll of two dice. To get a 2 you have to get a 1 on both dice. Obviously the probability of getting a 1 on either die is 1/6, since a die has six sides. The probability of joint events, like having a 1 on both dice, is found from the following rule.

To find the probability of several events all happening, multiply their probabilities.

The two probabilities to be multiplied in the case of the dice are 1/6, the probability of getting a 1 on the first die, and 1/6, the probability of getting a 1 on the second die. Multiplying 1/6 by 1/6 we get 1/36. If you prefer to use decimal fractions, then we could say the same thing by saying 0.167 times 0.167 equals 0.028. Thus we see that rolling a 2 (snake eyes) is a rare event that happens less than three times per 100 rolls.

There is something we have to keep in mind using this multiplication rule for joint events. The events have to be independent, i.e., one happening does not depend on the other happening. This certainly holds for the two dice, since the two dice are two different objects and one die doesn't know what the other one is doing.

To illustrate the calculation of probabilities for joint events, let's try a problem.

Problem 18

Andy Clinton bets on football parlay cards. To win he has to pick the winning team in four out of four football games. The parlay cards also give a "point spread" for each game that the favored team has to win by in order to win a bet on them. For instance, if Notre Dame is favored by 10 points over Michigan, and Andy bets on them, Notre Dame has to win by at least 11 points for Andy to win. For any other result he loses. By setting up these point spreads, the oddsmakers have made it a 0.5 probability of winning on any game, no matter which side Andy chooses. Many gamblers think they are smarter than the oddsmakers and can see an advantage for one side after the oddsmakers have applied the point spread to even things up. But the oddsmakers are doing it for a living, and the gamblers are amateur second guessers.

Andy bets $1. If he picks all four winners, he gets $10. What is his mathematical expectation?

Given:

Four bets must be won to win
Probability of winning each bet = 0.5
Amount of bet = $1
Payoff for winning = $10

Find:

Mathematical expectation.

Solution:

The four football games are four independent events and the probability of winning each one is 0.5. Multiplying the probabilities together we get the joint probability of winning all four games.

$$0.5 \times 0.5 \times 0.5 \times 0.5 = 0.0625$$

Note that Andy does not get his original $1 back, whether or not he wins, so the probability of losing it is 1.0. We can now find the mathematical expectation. It is the probability of winning, 0.0625, times the payoff for winning, $10, minus the probability of losing the initial $1 times the amount lost.

$$(0.0625 \times \$10.00) - (1.0 \times \$1.00) = -\$0.375$$

Answer:

Andy's mathematical expectation is −$0.375. On the average he will lose $0.375 cents for each $1 he bets.

Now that we have seen how these joint probabilities work, let's try another problem.

Problem 19

Anita Wells enters a crossword puzzle contest in the newspaper. The contest requires completing 10 words with tricky definitions by putting in a single letter for each. Of the 10 words, 6 have a choice among four letters and 4 have a choice among three letters. All choices are equally probable. If Anita gets all 10 right she wins $10,000. It costs her a 20-cent stamp to mail the entry in. What is her mathematical expectation?

Given:

Six words have choice of four letters
Four words have choice of three letters
Payoff for winning = $10,000
Ten correct out of ten necessary to win
Cost of entry =$0.20

Find:

Mathematical expectation.

Solution:

In order to win Anita has to have all 10 winning letters. Each one that has a choice of four has a probability of 1/4 of being right and each one that has a choice of three has a probability of 1/3 of being right. Then the joint probability can be found by multiplying together the six factors of 1/4 and the four factors of 1/3.

$$(1/4)^6 \times (1/3)^4 = 0.000003014$$

This is an extremely small probability of about three chances in a million.

The mathematical expectation is the probability of winning found above times the payoff for winning minus the probability of losing her stamp money (a certainty) times the cost of the stamp.

$$(0.000003014 \times \$10{,}000) - (1.0 \times \$0.20) = -\$0.1686.$$

This rounds off to $0.17.

Answer:

Her mathematical expectation is −$0.17. She loses an average of $0.17 with each entry.

If this chapter makes you more wary of contests and gambling, it has served one of its purposes. There are many traps in life with negative mathematical expectations. Since you can think with numbers you can calculate the mathematical expectations and avoid these traps.

4. THINKING BIG AND THINKING SMALL

Big Numbers

Did you ever hear some six-year old kids play a game called, "How high can you count?"

"I can count to a hundred," says Johnnie.

"I can count to a thousand," says Sally.

"Did you ever count that high?" asks Johnnie.

"No, but a thousand is higher than a hundred," says Sally.

"A million is higher than a thousand," says Billy.

"A billion is higher than that," says Sammy.

"I bet a zillion is more," says Lucy.

"No," says Albert. "Infinity is the highest number. It's bigger than anything."

"Did you ever count up to it?" asks Johnnie.

Albert was more or less right. The count of numbers is infinite. Whatever number one person names,

another can add 1 to it and come up with a larger number. The symbol for infinity is ∞ (the lazy 8). Infinity is a useful concept in theoretical scientific work, but for practical matters it is necessary to have names and notations for the full range of large numbers.

Let's review the more frequently used large numbers.

A million, which is written 1,000,000, is a thousand thousand.

A billion, which is written 1,000,000,000, is a thousand million. To visualize how large $1 billion is, imagine you were paid $100,000 per year for 10,000 years.

There is no such thing as a zillion. Sorry, Lucy.

The number which is a thousand billion is called a trillion. That's a 1 with 12 zeros after it. 1,000,000,000,000. There are not too many everyday things that come in trillions. The prices of all goods sold in the entire country in 1 year is about $3 trillion.

The number which is a thousand trillion is called a quadrillion. It's written 1,000,000,000,000,000. This number sometimes comes up in discussing how many units of energy are being used in the nation.

The other numbers, going up at a factor of 1000 each, are:

1,000,000,000,000,000,000—a quintillion;
1,000,000,000,000,000,000,000—a sextillion;
1,000,000,000,000,000,000,000,000—a septillion.

Typically these numbers are used most often in astronomy.

The number 2,835,621,928,031,726,124,532,418 is read as two septillion, eight hundred and thirty-five sextillion, six hundred and twenty-one quintillion, nine hundred and twenty-eight quadrillion, thirty-one trillion, seven hundred and twenty-six billion, one hundred and twenty-four million, five hundred and thirty-two thousand, four hundred and eighteen.

It can be seen from the above that it gets pretty cumbersome to write out very large numbers. To get around this problem scientific notation was devised.

Scientific Notation

Scientific notation involves using 10s with exponents (those little numbers above the line) instead of adding a lot of 0s. For example, in scientific notation

$$1000 = 10^3$$
$$1{,}000{,}000 = 10^6$$

As you can see, *the exponent tells you how many 0s appear after the 1.* In writing 1000 as 10^3 we are merely indicating that 1000 is $10 \times 10 \times 10$.

Similarly, we can write the large numbers we discussed in the previous section by using scientific notation.

$$1 \text{ billion} = 10^9$$
$$1 \text{ trillion} = 10^{12}$$
$$1 \text{ quadrillion} = 10^{15}$$
$$1 \text{ quintillion} = 10^{18}$$
$$1 \text{ sextillion} = 10^{21}$$
$$1 \text{ septillion} = 10^{24}$$

We can combine a 10 with an exponent with another number to write any number we wish in scientific notation. For instance, we can write

$$1{,}200{,}000 = 1.2 \times 10^6$$

As another example we can write

$$13{,}340{,}000{,}000 = 1.334 \times 10^{10}$$

In scientific notation the usual form is to have a number with one digit to the left of the decimal point multiplied by the 10 and its exponent. Thus if we wrote instead

$$13{,}340{,}000{,}000 = 13.34 \times 10^9$$

this is mathematically correct, but it is not in the usual form. Calculators that present large numbers in scientific notation generally use the form where the lead number has one digit in front of the decimal point.

Scientific notation can also be used for writing numbers less than 1 by using negative exponents. For instance,

$$1 \text{ tenth} = 0.1 = 1/10 = 10^{-1}$$

In scientific notation the negative exponent just means the number of 0s after the 1 in the bottom of the fraction. Thus in the above example, the exponent is -1 because there is one 0 after the 1 in the bottom of the fraction when the number is written as the fraction 1/10. You will also note that the 1 appears in the first decimal place.

Similarly, if the exponent is -2, we have

$$1 \text{ hundredth} = 0.01 = 1/100 = 10^{-2}$$

Again we see that the exponent corresponds to the number of 0s after the 1 in the bottom of the fraction when the number is written as 1/100.

The rule for placing the decimal point is to start out with 1.0 and move the decimal point the number of places shown by the exponent. Move the decimal point to the left for a negative exponent or to the right for a positive exponent. Thus 10^{-1} is 0.1 and 10^{-2} is 0.01. Note that for negative exponents the number of 0s to the left of the 1 is one less than the exponent.

Using negative exponents, it's easy to write very small numbers. For instance,

$$1 \text{ millionth} = 0.000001 = 10^{-6}$$

Multiplying or dividing by numbers in scientific notation is especially easy because it just involves moving the decimal point. For instance, in multiplying by 10^{3} move the decimal point three places to the right. In multiplying by 10^{-3} move the decimal point three places to the left.

Another way you can think of the negative exponent is that it shows dividing by a number multiplied by itself that many times. So 10^{-6} really means

$$\frac{1}{10 \times 10 \times 10 \times 10 \times 10 \times 10}$$

since 10 is multiplied by itself six times in the bottom of the fraction.

As we did in the scientific notation for big numbers, we can put another number in front of the 10 and its exponent. This allows us to write any small number using scientific notation. Here are some examples.

$$0.0000186 = 1.86 \times 10^{-5}$$
$$0.0000000003 = 3 \times 10^{-10}$$
$$0.582 = 5.82 \times 10^{-1}$$
$$0.0000000000000000042 = 4.2 \times 10^{-18}$$

It can be seen from the above examples that scientific notation is very useful in writing extremely small numbers.

Computations Using Scientific Notation

Some of the more expensive calculators are equipped to do computations involving numbers written in scientific notation. If the one you own doesn't have this feature, don't worry about it. The computations are very easy to do by hand, as you are about to see.

Let's try a multiplication first.

$$(2 \text{ x } 10^2) \times (4 \times 10^3) = 8 \times 10^5$$

This is the same as writing $200 \times 4000 = 800{,}000$, which you should be able to check on your calculator. To get the answer, *multiply the numbers preceding the 10s, and then add the exponents of the 10s to get the new exponent.* Thus multiplying the numbers preceding the 10s, we get $2 \times 4 = 8$. Adding the exponents of the 10s, we get $2 + 3 = 5$ as the new exponent. The answer is 8×10^5.

The rule is similar for division. To divide two numbers using scientific notation, *divide the numbers preceding the 10s, and then subtract the exponents of the 10s to get the new exponent.*

Let's try some sample problems and show how scientific notation is used.

Problem 20

Suppose it costs $20 billion for a program to send a man to Mars. What will be the average amount it will cost each of 230 million persons who live in the United States?

Given:

Cost = $20 billion
230 million persons in U.S.

Find:

Average cost per person.

Solution:

Write the numbers using scientific notation. The cost is $\$2.0 \times 10^{10}$. The population is 2.3×10^{8} people. To divide in scientific notation, divide the numbers preceding the 10s and then subtract the exponents.

$$\begin{aligned}(2.0 \times 10^{10}) \div (2.3 \times 10^{8}) &= (2.0 \div 2.3) \times 10^{2} \\ &= 0.8696 \times 10^{2} \\ &= 86.96\end{aligned}$$

Answer:

It costs each person $86.96.

You'll find many uses for scientific notation in examining large budgets. Here's another example.

Problem 21

A charity hopes to raise money in New York City, which has a population of 7.9 million people. How much money will they raise if they get an average contribution of only 0.015 cents ($0.00015) per person?

Given:

Population = 7.9 million
Average contribution = $0.00015

Find:

Total money collected.

Solution:

Express both numbers in scientific notation. The population is 7.9×10^6. The average contribution is 1.5×10^{-4} dollars. Multiplying the numbers preceding the 10s and adding exponents, we have

$$(7.9 \times 10^6) \times (\$1.5 \times 10^{-4}) = \$11.85 \times 10^2 = \$1185$$

Answer:

They will collect $1185.

Scientific notation is also useful for problems involving very large numbers of very small objects, such as the following.

Problem 22

A large steam shovel bucket can hold 22 million grains of sand. If each grain of sand weighs 0.00014 lbs., how many pounds can the bucket hold?

Given:

22 million grains in bucket
Each grain weighs 0.00014 lbs.

Find:

Weight of sand.

Solution:

Use scientific notation.

$$(22 \times 10^6) \times (1.4 \times 10^{-4}) = 30.8 \times 10^2 = 3080$$

Answer:

The bucket holds 3080 lbs.

This chapter has shown some examples of thinking with big and small numbers using scientific notation. You should now be able to handle any number easily, from the largest to the smallest.

5. MONEY NOW—MONEY LATER

Thinking with numbers helps you to plan ahead, so you'll have money now and money later too. Then the thing that happened to Dick Ugson won't happen to you.

Poor Dick Ugson

Dick Ugson had always wanted to own a car although he wasn't making very much money at his job as a street sweeper. One day he was walking by an auto showroom window, and he saw a new gold and silver Fishtail 8 inside. After much hesitation, he walked inside and immediately a smiling salesman greeted him and shook his hand.

"Hi, I'm Sly Slickson," said the salesman.

"I'm Dick Ugson. I was wondering if maybe I could look at the Fishtail 8, the gold and silver one I saw in the window."

"Of course. Of course. Sit right down inside. Feel the real vinyl seats."

Dick sat.

"Just $18,328 delivered," said Sly.

"But I only have $6200," said Dick, thinking of his bank account.

"That's all right," said Sly. "We'll take that as a down payment, and you can finance the rest. Here, sign on the dotted line," he said, whipping out a contract.

Dick signed.

He felt proud of his new gold and silver Fishtail 8 as he drove it out of the showroom later that afternoon. He also felt very strapped for cash, no longer having any money in the bank.

The first month Dick was surprised at how much the Fishtail 8 was costing him. The monthly payments were $250, and the insurance payments were another $50. He found that gas, oil, parking, and repairs were costing him another $100. Rent, food, and clothing continued to cost him $300 per month as they had before buying the car. He found that since he had the car, every cent of his $700 per month of take-home pay was used up immediately. There was no money left for incidental expenses and savings as there had been before. He hadn't thought about all the operating costs when he bought the car and plunked down all his savings.

In the second month he met Zelda. She was beautiful and was attracted to Dick because of his new Fishtail 8. She demanded that he take her to the most expensive restaurants where a man who drives a Fishtail 8 should be seen.

At the end of the second month, Dick didn't have enough money for his car payment.

The next month the auto finance company sent Dick a nasty letter. The month after that he suddenly found his car gone. It had been repossessed. Zelda disappeared just as quickly.

Poor Dick Ugson! He was a victim of "first-cost blindness." He had thought only about the down payment and had forgotten about the operating costs.

Unfortunately, a lot of people who should know better have the same problem as Dick in thinking with numbers over time. Even our elected representatives often fall into this trap, worrying only about the effect on this year's budget and ignoring next year's continuing costs.

In this chapter you'll learn how to plan ahead with numbers, so what happened to Dick Ugson doesn't happen to you.

Cash Flow

Money is easy to think about when it's here in the present. Money that is in the future is harder. One of the ideas that makes it a little easier to visualize money over time is cash flow. You simply add up everything that comes in and subtract everything that goes out. The mathematics involved is pretty easy. The difficult part is having the foresight and the honesty to plan for the operating expenses.

For instance, in the case of Dick Ugson, his expenses while having the car looked like this.

Monthly car payment	$250
Insurance payment	50
Gas, oil, repairs, and parking	100
Rent, food, and clothing	300
Total expenses	$700
Total income	$700

Cash flow = $700 − $700 = 0

Since Dick's take-home pay was $700, this left him with a cash flow of 0. This was a very bad position for Dick to get himself into. He should have known that he wouldn't be able to continue with zero cash flow and no savings to fall back on.

When Dick met Zelda, his expenses looked like this for the month.

Dates with Zelda	$220
Insurance payments	50
Gas, oil, repairs, and parking	100
Rent, food, and clothing	300
Monthly car payments	250
Total expenses	$920
Total income	$700

Cash flow = $700 − $920 = −$220

With negative cash flow, something had to go, so Dick didn't pay his car payment. This resulted in disaster.

If Dick had thought ahead about the cash flow, he could have avoided this situation. He could have bought an Economy 4, a smaller, less expensive car, for $5800. He then would have been able to pay cash and still have $400 left in the bank, thus avoiding the expensive monthly payments. He would also have saved a little money on insurance, gas, and parking because of the smaller car. His monthly expenses with the Economy 4 would have looked like this.

Insurance payment	$ 40
Gas, oil, repairs, and parking	80
Rent, food, and clothing	300
Total expenses	$420
Total income	$700

Cash flow = $700 − $420 = $280

With a positive cash flow of $280 per month, Dick could have built his savings up again. He also would have been ready for unexpected expenses such as illnesses or expensive girlfriends.

Now that you're familiar with the idea of cash flow, let's try a couple of problems.

Problem 23

Tony Jackson is thinking of buying a boat for his family. If he does, he will need to spend $50 per month on rent for an anchoring space for the boat,

$30 per month on rent for a garage to keep it in at home, $80 per month for gasoline and provisions for the boat, $100 for the license for the boat each year, and $380 for the insurance for the boat each year.

Besides the boat, Tony's family expenses are $380 per month for rent, $120 per month for food, $1100 per year for clothing, $100 per month for automobile operating expenses, $435 per year for automobile insurance, $800 per year for furniture, and $600 per year for other expenses.

Tony earns $800 per month as a department store clerk, and his wife, Josephine, earns an additional $900 per year through occasional babysitting. What is the family's monthly average cash flow with and without the purchase of the boat?

Given:

Boat anchoring = $50 per month
Boat garage rent = $30 per month
Boat gasoline and provisions = $80 per month
Boat license = $100 per year
Boat insurance = $380 per year
Rent = $380 per month
Food = $120 per month
Clothing = $1100 per year
Automobile operation = $100 per month
Automobile insurance = $435 per year
Furniture = $800 per year
Other expenses = $600 per year
Tony earns $800 per month
Josephine earns $900 per year

Find:

Monthly cash flow without boat.
Monthly cash flow with boat.

Solution:

First restate all the costs as monthly costs. That means those that were stated as annual costs have to be divided by 12.

Monthly boat license cost = $100 ÷ 12 = $8.33
Monthly boat insurance = $380 ÷ 12 = $31.67
Monthly clothing cost = $1100 ÷ 12 = $91.67
Monthly auto insurance = $435 ÷ 12 = $36.25
Monthly furniture cost = $800 ÷ 12 = $66.67
Monthly other expenses = $600 ÷ 12 = $50.00
Josephine's monthly earnings=
$900 ÷ 12 = $75.00

The monthly expenses without the boat are:

Rent	$380.00
Food	120.00
Clothing	91.67
Automobile operation	100.00
Automobile insurance	36.25
Furniture	66.67
Other expenses	50.00
Total expenses	$844.59

Tony's income	$800.00
Josephine's income	75.00
Total family income	$875.00

Cash flow without boat = $875.00 − $844.59 = $30.41

With the boat, Tony would have all the expenses he had before plus the boat expenses.

Rent	$380.00
Food	120.00
Clothing	91.67
Automobile operation	100.00
Automobile insurance	36.25
Furniture	66.67
Other expenses	50.00
Boat anchoring	50.00
Boat garage rent	30.00
Boat gasoline and provisions	80.00
Boat license	8.33
Boat insurance	31.67
Total expenses	$1044.59

As was shown above, Tony's total family income is $875.00.

$$\text{Cash flow} = \$875.00 - \$1044.59 = -\$169.59$$

Answer:

Cash flow = $30.41 per month without boat.
Cash flow = −$169.59 per month with boat.

After performing this cash flow calculation, Tony decided he couldn't afford to buy the boat. He is just barely breaking even on cash flow now. If he had the added expense of the boat, he would have a large negative cash flow and would be dipping into savings on a regular basis. That would mean that his savings wouldn't be there when he needed them for a new car or other large cash outlay. Tony wisely decided he would wait until he was making a lot more money before getting a boat.

Let's look at another problem. This one involves a business decision.

Problem 24

Steve Harrison has a job fixing coin-operated laundry machines for a big company. They pay him $15,600 per year. He hears about a friend of his who left the company and started his own coin laundry in a small store. From his friend, Steve learns that such a store can be rented for $300 per month. Rental of the coin-operated machines will run $60 per month, and various parts for the machines and supplies will run $30 per month. Taxes and licenses for a coin-operated laundry like this run $200 per year. He can expect to collect $600 in coins from the machines every week. What is the annual cash flow from the coin-operated laundry? How does it compare with his salary?

Given:

Present salary = $15,600 per year
Store rent = $300 per month
Machine rent = $60 per month

Parts and supplies = \$30 per month
Weekly collection = \$600
Taxes and license = \$200 per year

Find:

Annual cash flow from coin-operated laundry.
Comparison with present salary.

Solution:

First it is necessary to convert everything to annual figures.

Annual store rent = \$300 × 12 = \$3600
Annual machine rent = \$60 × 12 = \$720
Annual parts and supplies = \$30 × 12 = \$360
Annual collections = \$600 × 52 = \$31,200

We can now calculate annual expenses.

Store rent	\$3600
Parts and supplies	360
Machine rent	720
Taxes and licenses	200
Total expenses	\$4880
Total income	\$31,200

Cash flow = \$31,200 − \$4880 = \$26,320

The difference between this cash flow and Steve's present salary is

\$26,320 − \$15,600 = \$10,720

Answer:

Annual cash flow is \$26,320. This is \$10,720 more than Steve Harrison's present salary.

In this last example Steve Harrison found he could make a lot more money with his own store. Of course the key estimate was the amount of collections he would make each week. Luckily, he had his friend's store for comparison, and he had a pretty good idea what the numbers would be.

Time Payments

There are many times in life when we have to figure out time payments. These might come up for things like car payments, mortgage payments, or payments on a loan. Similarly, you may be receiving payments like salary checks, rent on your property, or dividends on your stock.

Analyzing time payments often involves calculations of interest. Long ago in grammar school you probably learned the formula for calculating interest as

$$I = Prt$$

where I is the amount of the interest payment, P is the principal (the amount of money borrowed), r is the interest rate (percent per year), and t is the time (years). When the letters are written together with no symbol between them, as in the above formula, it means multiplication. We'll be using this formula throughout the problems in this chapter and the next.

To show how the formula is used, let's look at a typical computation.

Suppose the amount borrowed is $5000, the interest rate is 18%, and the time period is 3 years. Then we have

$$I = Prt = (\$5000)\ (0.18)\ (3) = \$2700$$

If, on the other hand, we want to find the interest rate, then the formula is rearranged as

$$r = \frac{I}{Pt}$$

As an example, imagine that you are receiving monthly interest payments of $143.75 on $20,000 you have loaned to someone. You will get the $20,000 back in one lump sum at the end of 2 years. What interest rate are you receiving?

Interest rates are always quoted on an annual basis, so we'll only look at 12 months of payments in the interest calculation. From the interest formula

$$r = \frac{I}{Pt}$$

we have a 1 year period (t = 1), and 12 months of interest payments (I = 12 × $143.75 = $1725). Substituting this in the interest formula, we have

$$r = \frac{1725}{20{,}000 \times 1} = 0.08625$$

The interest rate you are getting is 8.625%.

Now suppose we change the repayment plan. Suppose instead of getting your money back in one lump sum at the end, you get it back in 24 equal payments, one with each month's interest payment. The payments are each

$$\$20{,}000 \div 24 = \$833.33$$

Then the total payment you get each month is

$$\$143.75 + \$833.33 = \$977.08$$

Are you getting a different interest rate under this new repayment plan? That's right, you sure are!

The thing to remember about interest is that it's interest on the *unpaid balance*. If you're getting part of your $20,000 back with each payment, then your unpaid balance is being reduced. You no longer have $20,000 out on loan. In fact, after the first 12 months, you only have $10,000 out on loan, since $10,000 has been paid back at that time.

$$\$833.33 \times 12 = \$10{,}000$$

In order to get an estimate of the interest, we will apply a rule of thumb made for situations like these where the principal is paid back in several payments. The rule is to *calculate the interest on the average unpaid balance*. The average unpaid balance is simply the sum of the beginning balance and the ending balance divided by 2.

For example, in the case we are talking about here, the average unpaid balance is

$$\frac{\$20{,}000 + 0}{2} = \$10{,}000$$

The annual interest rate is the money received for the year's 12 payments divided by the average unpaid balance of 10,000.

$$(\$143.75 \times 12) \div \$10{,}000 = 0.1725$$

The interest rate is thus 17.25%. Note that when part of the principal was paid back each month, the interest rate was double what it was when the principal was not paid back until the end. This shows the time value of money. For a given dollar amount of interest payment the interest rate is always higher if you're payng back part of the money sooner.

Because banks know about the time value of money, a repayment plan like the one we just discussed is not likely to be the kind given by a bank. This plan had \$143.75 as each month's interest and \$833.33 as each month's payment toward reducing the principal. The problem with such a repayment plan is that, because of the declining average unpaid balance, the interest rate would be much lower than 17.25% at the beginning and much higher than 17.25% at the end if each month's interest was \$143.75. For the first payment the average unpaid balance for the first month was \$20,000. The annual interest rate represented by a monthly interest payment of \$143.75 is

$$(\$143.75 \times 12) \div \$20{,}000 = 0.08625.$$

The interest rate is 8.625%. Thus the first payment of \$143.75 represents only half the interest of 17.25% that is being paid overall.

For the last payment the unpaid balance is only \$833.33. The interest rate on the last payment is

$$(\$143.75 \times 12) \div \$833.33 = 2.07$$

The last payment of \$143.75 represents 207% interest on the unpaid balance! Because of this situation, repayment schedules of loans have different proportions of the payment applied each month to interest and to paying back the principal. In the early payments, a larger proportion goes for payment of interest, and in later payments, a larger proportion goes for repayment of principal.

Let's try some sample problems in calculating interest.

Problem 25

Ann Dugan borrows $6500 from the bank at 10% to finish her college education. She agrees to pay it off in 3 years by sending the bank $207.64 per month for 36 months. In her first month's payment of $207.64, how much is interest and how much is going toward paying back the principal?

Given:

Amount borrowed is $6500
Payments are $207.64 monthly
Payment time is 3 years
Interest rate is 10%

Find:

Amount of first payment that is interest.

Amount of first payment applied toward paying back the principal.

Solution:

The average unpaid balance for the first month is the whole amount, $6500. The monthly interest on $6500 at 10% is

$$(\$6500 \times 0.10) \div 12 = \$54.17$$

The remainder of the first month's payment goes toward reducing the principal owed.

$$\$207.64 - \$54.17 = \$153.47$$

Answer:

Interest part of first payment is $54.17.
Repayment of principal in first payment is $153.47.

The key thing to remember in all these interest problems is always to first find the average unpaid balance. Then you know the amount the interest is being paid on. A lot of people overlook this point and incorrectly calculate the interest based on the entire amount borrowed. They get answers which are much too low. Let's do a problem illustrating this point.

Problem 26

Louis Stratton needs to borrow $270. Sam, the loanshark, offers to lend him the money to be paid back in three monthly payments of $100 each. Sam tells Louis that the interest rate is 11% since the three payments total $300 and

$$(\$300 - \$270) \div \$270 = 0.1111.$$

What is the real interest rate?

Given:

Amount borrowed = $270

Repayment of principal and interest by three payments of $100 each

Find:

Interest rate.

Solution:

First find the average unpaid balance by dividing the sum of the beginning and ending balances by 2.

$$(\$270 + 0) \div 2 = \$135$$

The amount of interest paid during the 3-month period is $30. Then the interest rate for the 3-month period is

$$\$30 \div 135 = 0.22$$

The 3-month interest rate is 22%.

However, interest rates are always expressed as annual rates, not 3-month rates. To get from 3 months to 12 months we have to multiply by (12 ÷ 3), which is 4.

$$0.22 \times 4 = 0.88$$

Answer:

The real interest rate is 88%.

The previous example has illustrated three important points in interest calculation: (1) base the interest calculation on the unpaid balance; (2) normalize all interest rates to an annual basis; and (3) never trust loan sharks' calculations.

The final problem of the chapter is also about unpaid balance.

Problem 27

Sidney Williams loans $1200 to his brother-in-law for 1 year. The agreement between them is that Sidney will get $100 of the money back each month for 11 months and then at the end of the 12th month, he will collect $200. What interest rate is Sidney receiving?

Given:

Principal = $1200
$100 per month paid for 11 months
$200 paid in last month

Find:

Interest rate.

Solution:

The average unpaid balance is found by summing the beginning and ending unpaid balances and dividing by 2.

$$(\$1200 + 0) \div 2 = \$600$$

The amount of interest was

$$11 \times \$100 + \$200 - \$1200 = \$100$$

Thus the interest rate is

$$\$100 \div \$600 = 0.167$$

Answer:

The interest rate is 16.7%.

This chapter has presented an introduction to interest problems. The next chapter will tell more about interest problems including how to look up the answers you need in interest tables.

6. TIME IS MONEY

Compound Interest

Everyone knows that the Indians sold Manhattan Island for $24. Very few people realize that if they had invested their money at compound interest, they would now be able to buy it back! Yes, if they had put their $24 into an investment paying 8% interest in 1626, then in 1976, 350 years later, they would have had $12,000,000,000,000. If you remember Chapter 4, you know this is $12 trillion, or in scientific notation $\$12 \times 10^{12}$. The 1976 assessed value of Manhattan was only about $20 billion. So the Indians would have had enough to buy Manhattan and pay all the 1.5 million people living there about $8 million each for relocation. How could all that money come out of $24? That's what you're going to find out in this chapter.

As you might remember, compound interest is what happens when you collect interest on interest. At the end of every interest period, the interest is added to the existing balance. We talked about something like this in Chapter 2, when we discussed percentage of growth.

For instance, let's suppose you have an investment of $1000 with an interest rate of 8% per year. Typically for savings accounts and savings certificates, this would be compounded quarterly rather than annually. The quarterly interest rate is

$$8\% \div 4 = 2\%$$

This means that every quarter your money grows by 2%.

After the first quarter 2% of $1000, which equals $20, is added to your account. Then your balance going into the second quarter is

$$\$1000 + \$20 = \$1020$$

The second quarter's interest is

$$\$1020 \times 0.02 = \$20.40$$

which is added to your balance. This gives you

$$\$1020 + \$20.40 = \$1040.40$$

as your balance going into the third quarter. The third quarter's interest is

$$\$1040.40 \times 0.02 = \$20.81$$

Adding this to the existing balance gives the balance going into the fourth quarter as

$$\$1040.40 + \$20.81 = \$1061.21$$

The fourth quarter's interest is

$$\$1061.21 \times 0.02 = \$21.22$$

Adding this to the balance you had at the beginning of the fourth quarter gives your balance at the end of the year.

$$\$1061.21 + \$21.22 = \$1082.43$$

You have added \$82.43 to your balance for the year because of compound interest.

Your equivalent annual interest rate for the year was

$$\$82.43 \div \$1000 = 0.08243$$

This rate of 8.243% is slightly greater than 8%, the quoted rate, since the quoted rate doesn't take the quarterly compounding into account. This goes to show you that if you're looking at two investments with the same quoted interest rate, the one with more frequent compounding will have a slightly higher yield. Some banks now offer daily compounding, and these give you even a bit more than quarterly compounding.

Now that you've seen how compound interest works, let's talk about an easier way to find the future value of an investment. The formula that does it is

$$F = P \times (1 + r)^n$$

In this formula F is the future worth of the investment, P is the principal, r is the interest rate for the period, and n is the number of periods. In the example we just discussed P was \$1000, r for each quarter was 0.02, and n was 4, since we found the future worth at the end of the fourth quarter.

$$\begin{aligned} F &= \$1000 \times (1 + 0.02)^4 \\ &= \$1000 \times 1.02 \times 1.02 \times 1.02 \times 1.02 \\ &= \$1082.43 \end{aligned}$$

This is the same answer as we had before, but we got it in a shorter way.

Using Tables to Find Compound Interest

There is an even easier way to do this calculation by using the tables in the back of this book. Turn to the table that says 2% at the top, and we'll talk about how to do it.

Look at the table and you'll see that the first column says n at the top and the second one says U.

Don't worry about the other columns yet; you'll learn about them later in this chapter.

Look down the first column and under n you'll see the number 4. Right next to it in the second column you'll see the number 1.082. (If you don't see it you're probably in the wrong table.) The number 1.082 is the one we need to calculate the future worth of an investment compounded four times at 2%. Applying the number from the table to our examples gives

$$\$1000 \times 1.082 = \$1082$$

This is the same answer that we got before, except that the answer from the table doesn't give the amount of cents. Still, the table answer is accurate enough for most purposes and a lot easier to calculate.

Let's look at the table again, and this time let's see what your $1000 investment will be worth in 10 years or 40 quarters. What number do you see opposite 40 in the table?

That's right. The number 2.208 is opposite 40. The answer then is

$$F = \$1000 \times 2.208 = \$2208$$

After 10 years your $1000 investment has more than doubled and is now worth $2208.

Now that you know how to use the tables, let's try some sample problems.

Problem 28

Dorothy Washington loans $2500 to her son for 8 years. They agree that he will return the money at the end of that time with 6% interest compounded annually. How much money does he give her when the loan becomes due?

Given:

Principal = $2500
Interest rate = 6%
Time = 8 years
Compounding is annual

Find:

Future worth.

Solution:

Use the table at the back of the book that says 6% at the top. Look down the n column and find 8. The number opposite 8 in the U column is 1.594. The amount due in 8 years is

$$\$2500 \times 1.594 = \$3985$$

Answer:

Her son returns $3985.

Problem 29

How much money would Dorothy Washington receive from her son if they had agreed to quarterly compounding instead of annual? Again, as in Problem 28, it's $2500 loaned at 6% interest for 8 years.

Given:

Principal = $2500
Interest rate = 6%
Time = 8 years
Compounding is quarterly

Find:

Future worth.

Solution:

The compounding is quarterly. The interest rate each quarter is

$$6\% \div 4 = 1.5\%.$$

Therefore we use the 1.5% table in the back of the book. The number of quarters is

$$8 \times 4 = 32$$

The number we want is in the U column opposite 32 in the 1.5% table. It is 1.610.

Answer:

The future worth is F = $2500 × 1.610 = $4025.

In comparing the answers to Problems 28 and 29 we see that quarterly compounding does make a difference. The future value with quarterly compounding is $4025 and with annual compounding it's $3985. That's a difference of $40.

Now that you're used to the tables, let's go back and try the problem we talked about at the beginning of the chapter, the interest on the Manhattan Indians' $24.

Problem 30

What would be the worth at the time of the 1976 Bicentennial of the Manhattan Indians' $24 if they had invested it at 8%, compounded annually, when they received it in 1626?

Given:

Principal = $24
Interest rate = 8%
Beginning date is 1626
Ending date is 1976
Compounding is annual

Find:

Worth of investment in 1976.

Solution:

The number of years of the investment is

$$1976 - 1626 = 350$$

Obviously we want to use the 8% table for this calculation. But you'll notice that the table only goes to 100 periods, not 350. In order to do the calculation for 350 years, we'll have to multiply the factor for 100 years by itself three times and then multiply that by the factor for 50 years. Looking in the table, you'll find that the factor for n = 100 is 2199.761 at the bottom of the U column. Expressing it in scientific notation it is 2.199761×10^3. For n = 50 the factor is 46.902. Thus we have

$$
\begin{aligned}
F &= \$24 \times (2.199761 \times 10^3)^3 \times 46.902 \\
&= \$24 \times (10.645 \times 10^9) \times 46.902 \\
&= \$11{,}982 \times 10^9 \\
&= \$11.982 \times 10^{12}
\end{aligned}
$$

This is roughly $12 trillion.

Answer:

Worth of investment in 1976 = $12 trillion.

Besides problems in investments like those we have solved thus far, compounding can also be used for calculating the effects of inflation on prices, as in the next problem.

Problem 31

A can of peaches now costs $0.87. How much will it cost in 10 years if inflation is 10% per year?

Given:

Price of can of peaches	=	$0.87
Inflation rate	=	10%
n	=	10 years

Find:

Price of can of peaches in 10 years.

Solution:

Looking at the 10% table in the back of the book, we find that the value of U for n = 10 years is 2.594.

Solving for future value, we have

$$F = \$0.87 \times 2.594 = \$2.26$$

Answer:

The price of the can of peaches in 10 years will be $2.26.

Thus we see how dangerous inflation is. If we had money in the bank at 5%, it would only be growing half as fast as the inflation rate, and we would be able to buy less later than we could now, despite the fact that our money had grown. Investors realize this and in times of high inflation they have demanded very high interest rates.

Besides using the tables for finding future worth from present principal, you can also use them for the opposite problem—finding present principal from future worth.

For instance, in the example we discussed at the beginning of the chapter, the future worth at the end of four quarters was $1082. The interest rate per quarter was 2% since it was 8% annual interest. Suppose we want to find the present principal from the future worth. We can use the third column, marked V, in the 2% table. In this column, we see the entry 0.9238 opposite n = 4. We can use this to find the principal as follows.

$$P = \$1082 \times 0.9238 = \$1000$$

This answer checks with what we already know, that the original principal was $1000.

Now that we know how to use the V column of the table, let's try two sample problems with it.

Problem 32

Minnie Andrews wants to put some money in the bank now so that she can replace her present car in 6 years when it wears out. She will be putting the money in a savings certificate yielding 7% compounded semiannually. She expects that the car she will want to buy in 6 years will cost $6000 then. How much money should she put in the bank now to cover the car?

Given:

Future worth = $6000
Interest rate = 7%
Time = 6 years
Compounding is semiannual

Find:

Present principal.

Solution:

Since the compounding is semiannual, there are two compoundings per year or

$$6 \times 2 = 12$$

compoundings in 6 years. Each compounding is at a rate of

$$7\% \div 2 = 3.5\%$$

Therefore we will use the 3.5% table. In the 3.5% table in the V column, we see that the entry opposite 12 is 0.6618. The calculation is then

$$P = \$6000 \times 0.6618 = \$3970.80$$

Answer:

She should put $3970.80 in the bank.

Problem 33

Anthony Warren learns about an investment that promises to return $10,000 in 10 years if he puts in $5000 now. What interest rate would this investment be providing under annual compounding?

Given:

F = $10,000
P = $5,000
Time = 10 years
Compounding is annual

Find:

Interest rate.

Solution:

This is a little bit different from the previous problems since we are now trying to find the interest rate. In previous problems it was given. Like the previous problems this is going to involve using the tables in the back of the book. The thing we're looking for in the tables is a U factor for $10,000 of future worth coming from $5,000 of present principal. That factor is

$$U = \$10{,}000 \div \$5{,}000 = 2.0$$

We should find 2.0, or some number close to it, in the U column of one of the tables opposite n = 10.

Scanning the various tables, we find an entry of 1.967 in the 7% table and an entry of 2.159 in the 8% table. From these numbers we can see that the interest rate must be a little more than 7%.

Answer:

Interest rate is slightly over 7%.

This checks with what we learned earlier in Chapter 2 about the Rule of 70. The example showed a doubling of the money in 10 years, so we would expect the interest rate to be

$$70 \div 10 = 7$$

The Rule of 70 also gives 7% as the interest rate for this case. Although the Rule of 70 is handy to use, it can only be used in cases like this where there is a doubling. Otherwise, you have to use the tables.

Annuities

An annuity is a repeated payment. Examples of annuities are car payments or mortgage payments. You can figure out how much these payments should be by using the tables in the back of this book. In the remainder of this chapter we're going to see how it's done.

Let's begin by working out an annuity problem without the table so you can see what's involved. Suppose that you put $200 in the bank at the end of every quarter, and the bank pays 6% interest compounded quarterly. How much money do you have in the bank after five quarters?

Well, you'll have the five payments you put in plus the interest on them. Let's do it a step at a time.

At the end of the first quarter you just have your first payment of $200. You've just put it in so it hasn't collected any interest yet.

At the end of the second quarter you have the second \$200 plus the first \$200 plus one quarter's interest on the first \$200. Since the interest rate is 6% annually, it's 1.5% for one quarter. So at the end of the second quarter you have

$$\$200 + \$200 + (\$200 \times 0.015) = \$403$$

At the end of the third quarter you'll have the \$200 you just put in, plus the \$403 from before, plus interest on the \$403.

$$\$200 + \$403 + (\$403 \times 0.015) = \$609.05$$

Similarly at the end of the fourth quarter there's the \$200 you just put in, plus the \$609.05, plus interest on the \$609.05.

$$\$200 + \$609.05 + (\$609.05 \times 0.015) = \$818.19$$

Finally at the end of the fifth quarter, you'll have the \$200 you just put in, plus the \$818.19, plus the interest on \$818.19.

$$\$200 + \$818.19 + (\$818.19 \times 0.015) = \$1030.46$$

Now that you see how the process works, let's see how you can find the answer a lot faster using the tables at the back of the book. This time you're going to be using column Y, the second from the last one.

Turn to the table in the back of the book for 1.5%. Now look in the Y column for $n = 5$. What number do you see? That's right, it's 5.152. To find the future worth F, we multiply the annuity A by the factor you just looked up Y. The formula is

$$F = AY$$

Then the answer to our problem is

$$F = \$200 \times 5.152 = \$1030.40$$

This is the same answer that we got the long way before, except for the last cents digit. This is close enough for all practical purposes.

Now that you know how to use the Y column of the table to find the future worth of an annuity, let's try a few problems.

Problem 34

Nathan Green puts $150 in the bank at the end of every quarter for 10 years. How much money does he have at the end of 10 years if the bank pays 5% compounded quarterly?

Given:

Annuity = $150
Deposits at end of each quarter
Interest rate = 5%, compounded quarterly
Duration = 10 years

Find:

Future worth at end of 10 years.

Solution:

Since the interest rate is 5% annually, it's 1.25% for one quarter.

$$5\% \div 4 = 1.25\%$$

There are 40 quarters in 10 years.
Looking in the table under column Y for n = 40, we see the factor we need is 51.490.

$$F = \$150 \times 51.490 = \$7723.50$$

Answer:

After 10 years he has $7723.50 in the bank.

Problem 35

Blanche Parker puts $225 in a savings certificate account at the end of every quarter. The account has an interest rate of 7% compounded quarterly. How much money will she have in the account at the end of 15 years?

Given:

Annuity = $225
Deposits at end of each quarter
Interest rate = 7%, compounded quarterly
Duration = 15 years

Find:

Future worth at end of 15 years.

Solution:

This is just like the previous problem, with different numbers. The interest rate per quarter is

$$7\% \div 4 = 1.75\%.$$

The number of quarters is

$$15 \times 4 = 60$$

Looking in the 1.75% table, we find that the entry in column Y opposite $n = 60$ is 104.675.

Then the future worth is

$$F = \$225 \times 104.675 = \$23{,}551.88.$$

Answer:

Blanche Parker will have $23,551.88 in the account at the end of 15 years.

Problem 36

Milton Edwards is paying off an installment at 1.5% per month, compounded monthly. He puts in $35 at the end of every month. How much money will his payments be worth to the finance company at the end of 2 years?

Given:

Annuity = $35
Deposits at the end of each month
Interest rate = 1.5% per month, compounded monthly
Duration = 2 years

Find:

Future worth at end of 2 years.

Solution:

The interest rate is 1.5% per month, so we will be using the 1.5% interest table. The number of months in 2 years is

$$2 \times 12 = 24$$

Therefore we need to look up Y in the 1.5% table for

n = 24. The factor we find in the table is 28.634. Then we find the future worth.

$$F = \$35.00 \times 28.634 = \$1002.19$$

Answer:

Milton Edwards' payments will be worth $1002.19 to the finance company at the end of two years.

The above answer may surprise some people since all this money came from $35 per month. Actually this is what we'd expect for a lot of payments (24) and a high interest rate (1.5% per month equals 18% per year). The factor Y = 28.634 in the above example means that the interest converts 24 payments into the equivalent of 28.634.

Problem 37

Joanne Strong is visited by a life insurance salesman. The salesman tells her that if she pays a premium of $400 at the end of each year, at age 65 she will collect $30,000. She is 30 years old now and has no dependents. The salesman tells her she should buy the life insurance as a way to save for her retirement. What interest rate is the insurance company providing on her money, if she reaches age 65?

Given:

A = $400
F = $30,000
Age now = 30 years
Age when money collected = 65 years

Find:

Interest rate.

Solution:

In problems like this one where you have to find the interest rate, you first calculate the factor, and then look to see which table it's in. First let's calculate the factor Y.

$$
\begin{aligned}
Y &= F \div A \\
&= \$30{,}000 \div \$400 \\
&= 75.0
\end{aligned}
$$

The number of years is

$$65 - 30 = 35$$

Next we have to look through the tables to see which table has a number close to 75.0 in the Y column in the n = 35 row.

Looking through the table for 4%, we find Y = 73.652 for n = 35, and in the table for 4.5% we find Y = 81.497 for n = 35. Our Y value of 75.0 is between these, so the interest rate is between 4% and 4.5%. Since the Y value is closer to the 4% Y value, an accurate description would be to say the interest rate is slightly more than 4%.

Answer:

The insurance company is providing her with an interest rate slightly over 4%.

This is a very low interest rate compared to rates she might be getting from bonds or savings banks. Since she doesn't need the insurance feature of providing money to some dependent in case of her death, she is better off investing her money elsewhere. The only reason she might have for putting her money in life insurance instead of a bank is that she might not have the willpower to save if she didn't get a bill every year from the insurance company. This is an expensive price to pay for willpower.

Problem 38

Simon Walters, who is 22 years old, is thinking about buying life insurance. He is trying to decide whether to buy term insurance for 20 years or a 20-year endowment policy. He doesn't think he will need any more insurance if he lives through the next 20 years, since by then his two children will be grown. He is trying to figure out which policy to choose of the two that have been offered him by different companies.

If he buys the 20-year endowment policy, his annual premium will be $600. The policy provides that he will collect $20,000 in 20 years at age 42. What interest rate will he have received?

If he buys a $20,000 term insurance policy, the rate will be $80 per year. If he buys term insurance, Simon expects to save the remaining $520 he would have spent on insurance premiums by annually putting it into 7% certificates of deposit at his bank. How much money will he have through his savings at the end of 20 years?

Which is the better choice for Simon Walters—the 20-year endowment policy or the term insurance coupled with savings?

Given:

(For the 20-year endowment policy)

A = $600
F = $20,000
n = 20 years

(For the term insurance plus savings)

Annual premium = $80
Value of policy = $20,000
Annual investment = $520
Interest rate = 7%
n = 20 years

Find:

Interest rate on 20-year endowment policy.
Future worth of savings with term-insurance-plus-savings plan.

Solution:

The 20-year endowment policy calculation is just like the previous problem with different numbers.

$$Y = F \div A$$
$$= \$20,000 \div \$600$$
$$= 33.333$$

In the $n = 20$ rows of the tables, we find $Y = 33.066$ in the 5% table and $Y = 34.868$ in the 5.5% table.

Answer:

The 20-year endowment policy is giving him just over 5%.

For the term-insurance-plus-savings choice, the annual premiums paid are not returned. However, they are much lower than the premiums on the 20-year endowment policy since Simon is not paying for a savings feature. Thus he is able to make annual savings of

$$A = \$520$$

The period is

$$n = 20 \text{ years}$$

To find the future worth of these annual savings at 7%, we use the formula

$$F = A\,Y$$

In the 7% table for $n = 20$ years, we find $Y = 40.995$. Thus

$$\begin{aligned} F &= 520 \times 40.995 \\ &= \$21{,}317.40 \end{aligned}$$

Answer:

The savings will be worth \$21,317.40 at the end of 20 years.

The term insurance plus savings is Simon Walters' better choice.

It turned out that Simon did better by putting his money in a certificate of deposit and getting a better interest rate, rather than saving via insurance. Instead of buying life insurance with a savings feature, you will generally do better by buying term insurance and investing the money you saved from having a lower premium, provided you have the willpower to save.

Mortgages

Most Americans buy a house at some time during their lives, and this usually means taking out a mortgage. In this section we're going to talk about how to do mortgage calculations using the interest tables at the back of the book.

Let's suppose you want to buy a $40,000 house. You are putting down an $8000 down payment so the remainder for which you have to get a mortgage is

$$\$40{,}000 - \$8000 = \$32{,}000$$

The mortgage that you get has a 10% interest rate and runs for 20 years. How much are the annual payments?

Let's turn to the 10% table at the back of the book. The column we're going to use this time is the one labeled X. Look down column X until you see the n = 20 row. The entry you see there is 0.11746. The annual payment A is then found from the formula

$$A = PX$$

where P is the principal, $32,000 for this case, and X is the number from the X column of the table.

$$\begin{aligned} A &= \$32{,}000 \times 0.11746 \\ &= \$3758.72 \end{aligned}$$

The annual payment, which consists of interest and paying off part of the principal, is $3758.72.

It is pretty easy to find mortgage payments using the interest tables. Let's try a few problems.

Problem 39

Marvin Wellington is taking out a mortgage on his new house to cover the balance that he owes of $63,000. He is getting a 15-year mortgage with payments made quarterly and quarterly compounding at an annual interest rate of 12%. How much are his quarterly payments?

Given:
P = $63,000
Interest rate 12%, compounded quarterly
Payments are quarterly
Duration = 15 years

Find:
Quarterly payment.

Solution:
The number of quarterly payments is

$$15 \times 4 = 60$$

The quarterly interest rate is

$$12\% \div 4 = 3\%$$

Looking in the 3% table for n = 60 we find X = 0.03613.

Then the annuity we're looking for is

$$\begin{aligned} A &= \$63{,}000 \times 0.03613 \\ &= \$2276.19 \end{aligned}$$

Answer:
His quarterly payment is $2276.19.

Problem 40

Florence Wilson buys a condominium for $53,000. Her down payment is $11,000. The remainder is covered by a 25-year mortgage at 10% with payments made semiannually and semiannual compounding. How much are the payments?

Given:
Value of condominium = $53,000
Down payment = $11,000
Interest rate = 10%
Payments are semiannual
Compounding is semiannual
Duration = 25 years

Find:
Amount of payment.

Solution:

The amount covered by the mortgage is

$$\$53{,}000 - \$11{,}000 = \$42{,}000$$

The number of payments is

$$25 \times 2 = 50$$

The semiannual interest rate is

$$10\% \div 2 = 5\%$$

Looking in the 5% interest table for $n = 50$, we find $X = 0.05478$. The payment is

$$A = \$42{,}000 \times 0.05478$$
$$= \$2300.76$$

Answer:

Her semiannual payments are $2300.76.

Now that you have the idea, let's try some mortgage problems where you are finding the interest rate. As in the previous problems where you found the interest rate, you're going to have to flip through the tables to find which table has the factor value closest to the one you're looking for.

Problem 41

Homer Weatherly has a 20-year $45,000 mortgage. His annual payments are $6000. What interest rate is he paying?

Given:

$P = \$45{,}000$
$A = \$6000$
Payments are annual
Duration is 20 years

Find:

Interest rate.

Solution:

As in previous problems of finding an interest rate, we're going to have to find a factor first. In this case it's X.

$$X = A \div P$$
$$= \$6000 \div \$45{,}000$$
$$= 0.13333$$

Looking through the various tables in the n = 20 row, we find X = 0.13388 in the 12% table.

Answer:
The interest rate is 12%.

Problem 42

Andy Winston has a 20-year $30,000 mortgage on his condominium. Interest on it is compounded semiannually and his semiannual payments are each $1600. What interest rate is he paying?

Given:
P = $30,000
A = $1600, semiannually
Compounding is semiannual
Duration = 20 years

Find:
Interest rate.

Solution:
This is very much like the previous problem, except that we have to solve first for the semiannual interest rate. To look it up we're going to have to find X.

$$X = A \div P$$
$$= \$1600 \div \$30{,}000$$
$$= 0.05333$$

Since we have a 20-year mortgage with payments and compounding semiannual

$$n = 20 \times 2 = 40$$

Thus we must look through the n = 40 rows of the table to try to find an X value close to 0.05333.

In the 4.5% table we find X = 0.05434 for n = 40. So the semiannual interest rate is close to 4.5%. The annual rate is twice this.

$$4.5\% \times 2 = 9\%$$

Answer:

He is paying 9%.

Now we know how to do mortgage problems both ways, solving for the annuity or solving for the interest.

Retirement Plans

You've probably been wondering what we're going to do with the other two columns in the tables, W and Z. Well, you're going to learn about them now. As it turns out, these two columns are useful, among other things, for doing calculations on retirement plans.

For instance, suppose you want to put a sum of money in the bank at 5% at the end of every year so that in 30 years you'll have $40,000. How much money should you put in every year?

The answer to this comes from the W column of the 5% table and the formula involved is

$$A = FW$$

where we're solving for the annuity A. Looking in the W column of the 5% table for $n = 30$, we find $W = 0.01505$. Then we have

$$\begin{aligned} A &= \$40{,}000 \times 0.01505 \\ &= \$602 \end{aligned}$$

The answer is that you should put $602 in the bank at the end of every year to have your $40,000 at the end of 30 years.

The remaining column of the table is the Z column. Let's learn how to use it. Suppose you want to set up a retirement fund in the bank that allows you to draw $7000 per year for 10 years from it, before it's all gone. How much do you have to have in the bank at 5% in order to do this?

As you probably guessed, we're going to use the Z column in the table to solve this. The formula this time is

$$P = AZ$$

Looking in the 5% table for $n = 10$, we find $Z = 7.722$. Then the answer is

$$P = \$7000 \times 7.722$$
$$= \$54{,}054$$

It's pretty easy to use the W and Z columns since we've already learned about the other columns. Let's try a few problems.

Problem 43

Stephanie Swift is trying to save $60,000 over the next 25 years for her retirement. Her bank pays 5.5% interest. How much money does she have to put in the bank at the end of each year?

Given:

F = $60,000
Interest rate = 5.5%
Duration = 25 years

Find:

Annuity she must deposit.

Solution:

Use the W column of the 5.5% interest table. For $n = 25$ we find $W = 0.01955$. Then we have

$$A = FW$$
$$= \$60{,}000 \times 0.01955$$
$$= \$1173$$

Answer:

She must deposit $1173 at the end of each year.

Problem 44

Mathilda Heatherby wants to retire with enough money in the bank at 6% interest so that she can draw $2000 per quarter for 15 years before it's all gone. How much money does she need in the bank? The bank compounds quarterly.

Given:

A = $2000, per quarter

Duration = 15 years
Compounding is quarterly

Find:
Principal in the bank.

Solution:
The interest rate per quarter is

$$6\% \div 4 = 1.5\%$$

The number of quarters involved is

$$15 \times 4 = 60$$

Since this is a problem where we're given A and have to find P, we need to use the Z column of the 1.5% table. Looking there we find Z = 39.380 for n = 60. Then we have

$$\begin{aligned} P &= AZ \\ &= \$2000 \times 39.380 \\ &= \$78{,}760 \end{aligned}$$

Answer:
She has to have $78,760 in the bank.

Problem 45

James Davidson is planning for his retirement, 25 years from now. He wants to have enough money in 7% savings certificates which are compounded quarterly so that he can draw out $1500 per quarter and the money will last 10 years. How much money will he have to have at retirement time?

In order to get together the money he needs for retirement time, he plans to deposit a sum of money at 7% at the end of each quarter between now and then. How much does he have to deposit?

Given:
Interest rate = 7%, compounded quarterly
Duration to retirement = 25 years
Duration for retirement money to last = 10 years
A = $1500 per quarter, during retirement

Find:

P, principal necessary to provide retirement money.

A, annuity necessary to accumulate retirement fund.

Solution:

This is a double-barreled problem. The first thing we have to do is find the principal P that will provide the 10 years of retirement money.

Since this will be 10 years of quarterly payments, the number of payments is

$$10 \times 4 = 40$$

and the quarterly interest rate is

$$7\% \div 4 = 1.75\%$$

We will be using the 1.75% interest table for this problem. The first thing we have to find is the Z value for $n = 40$. Looking in the table we find $Z = 28.594$. Now we can solve the first part.

$$\begin{aligned} P &= AZ \\ &= \$1500 \times 28.594 \\ &= \$42{,}891 \end{aligned}$$

The amount James Davidson will need to have in the bank at retirement is $42,891.

To see how he's going to save it, we're going to have to look in the table under the number of quarters corresponding to 25 years.

$$n = 25 \times 4 = 100$$

The factor we want this time is W. That's the one that will get us from the future worth of $42,891 to the required annuity.

$$A = FW$$

Looking in the 1.75% table for $n = 100$, we find $W = 0.00375$. Then the required annuity is

$$\begin{aligned} A &= \$42{,}891 \times 0.00375 \\ &= \$160.84 \end{aligned}$$

Answer:

He will have to have $42,891 at retirement time.

He will have to deposit $160.84 each quarter to accumulate the money.

After going through the problems in this chapter you should be able to handle any interest calculation that's likely to come up. It's really pretty easy to do compound interest calculations if you have a set of interest tables and a calculator.

7. YOU CAN'T HAVE YOUR CAKE AND EAT IT TOO!

What Is a Trade-Off?

A trade-off is a situation where you have to give up something to get something else. It's very much like a budget. Trade-offs come up all the time in life, and you have to know how to think with numbers to deal with them. If you don't realize you're in a trade-off situation, and you try to have your cake and eat it too, you may wind up hungry with no cake either!

For instance, look what happened to Percival Ugmeister. Percival was taking a business trip to Hawaii and Alaska. So he got out all of his Hawaiian shirts and Bermuda shorts and his beach umbrella and packed them into his big suitcase. Then he remembered he was also going to Alaska, so he packed a muffler into the suitcase. He would have packed more cold weather clothing except that the suitcase was now full, and he remembered something about the airlines having a 40-lb. limit on baggage.

The bag seemed pretty heavy as Percival dragged it to the airport. Sure enough, when he arrived at the airline check-in counter, his baggage weighed in at 58 lbs. He was 18 lbs. overweight. At $1 a pound excess baggage charge, Percival had to pay the clerk $18 before checking his bag through.

Percival had a wonderful time in Hawaii. He put up his beach umbrella, sat under it in his Hawaiian shirt and Bermuda shorts, and watched the girls go by in their bikinis. As it turned out, Percival had taken ten times as many Hawaiian shirts and Bermuda shorts outfits as he needed for his 2-day stay in Hawaii. He just packed them back into the suitcase without getting a chance to wear them. He felt especially conscious of his oversupply of summer clothes as he paid another $18 overweight charge at the airline check-in counter at the Honolulu airport. Then he boarded his plane for Alaska.

Upon arriving in Alaska, the first thing he realized was that he was very cold. He still had on his summer-weight suit, and the only winter-weight thing he had in his suitcase was his muffler. It suddenly dawned on Percival that he would be staying 12 days in Alaska and he would need more winter clothes. Freezing, he ran to the nearest department store and bought a winter suit, a fur hat, a sweater, an overcoat, wool socks, galoshes, and long underwear. The bill came to $1388.92. He didn't think it would be so expensive, until he remembered that everything had to be imported from the "lower 48" states. The thing that really bugged him was that he had all these winter clothes at home; he just hadn't brought them.

Discovering how high the prices were, Percival didn't buy any more clothes, though he was cold all the time. He just wore the same suit day after day and shivered. Toward the end of the 12 days, some of his business associates asked why he hadn't changed suits and confided to him that he was becoming a bit smelly.

Percival was glad to go to the airport to leave for home. At the check-in counter he had a large bundle of clothing that would no longer fit in the suitcase, because of his recent purchases. He had to pay an overweight baggage charge of $41 this time. As soon as he got home, Percival Ugmeister came down with pneumonia.

Percival's problem was that he had ignored the trade-offs! He obviously should have taken more winter clothes and left the beach umbrella and most of the summer clothes behind.

Since he had 30 lbs. of room left after packing his necessities, he should have figured out how he wanted to split the 30 lbs. between winter clothes and summer clothes. For instance, if he took 5 lbs. of winter clothes, that would leave him 25 lbs. for summer clothes.

$$30 - 5 = 25$$

If he took 10 lbs. of winter clothes, he could take 20 lbs. of summer clothes.

$$30 - 10 = 20$$

If he took 15 lbs. of winter clothes, he could take 15 lbs. of summer clothes.

$$30 - 15 = 15$$

Or if he took 20 lbs. of winter clothes, he could take 10 lbs. of summer clothes.

$$30 - 20 = 10$$

If he took 25 lbs. of winter clothes, he could take 5 lbs. of summer clothes.

$$30 - 25 = 5$$

The other two possibilities are taking all summer clothes or all winter clothes.

All of this can be summed up in a table listing the trade-offs between the weight of summer clothes that he could take and the weight of winter clothes.

Pounds of Winter Clothes	Pounds of Summer Clothes
0	30
5	25
10	20
15	15
20	10
25	5
30	0

Which of these weight combinations he picked would require some judgment. Since Percival was staying only 2 days in Hawaii compared to 12 in Alaska, he should have estimated that he would need five or six times as much weight of cold weather clothing. He might have picked a combination of 5 lbs. of summer clothes and 25 lbs. of winter clothes if he had figured out the trade-off.

The first thing you have to do in a trade-off situation is to realize that a trade-off exists. The next thing you have to do is to calculate a table listing the choices, like the one we just looked at. By examining the table of possible choices you can then make an intelligent decision on what you should do.

Now that you have the idea, let's try a couple of examples of trade-off problems.

Problem 46

Dolores Grand is packing a box of books and clothing to go to her son, Oscar, at summer camp. Since she is sending the box parcel post, she is allowed to ship no more than 70 lbs. Make a table showing the trade-offs for 10-lb. increments of books versus clothing.

Given:

Weight limit is 70 lbs.
Increments of weight are 10 lbs.
Trade-off choice is between books
and clothing

Find:

Table of trade-offs.

Solution:

This is similar to the Percival Ugmeister example. The increments are 10 lbs., so we will look at 0, 10, 20, 30, 40, 50, 60, and 70 lbs. of books. Calculating the weight left for clothing in each case, we find the following.

For 0 lb. of books
$70 - 0 = 70$ lbs. of clothing

For 10 lbs. of books
$70 - 10 = 60$ lbs. of clothing

For 20 lbs. of books
$70 - 20 = 50$ lbs. of clothing

For 30 lbs. of books
$70 - 30 = 40$ lbs. of clothing

For 40 lbs. of books
$70 - 40 = 30$ lbs. of clothing

For 50 lbs. of books
$70 - 50 = 20$ lbs. of clothing

For 60 lbs. of books
$70 - 60 = 10$ lbs. of clothing

For 70 lbs. of books
$70 - 70 = 0$ lbs. of clothing

Answer:

The trade-off table is as follows:

Pounds of Books	*Pounds of Clothing*
0	*70*
10	*60*
20	*50*
30	*40*
40	*30*
50	*20*
60	*10*
70	*0*

Having developed the trade-off table, Dolores Grand could then pick a choice from it. A letter from Oscar might give her some idea of which to send more of — books or clothing.

Problem 47

Edith Bartlet and Felicia McCarthy are in charge of picking out the door prizes for the office Christmas party. They are given a fund of $170 to work with. There are 97 people coming to the party. What will be the value of each door prize if they decide to give one to each person? What if they raffle them off and give out only 20? How about 10? 5? 4? 3? 2? 1? Make a trade-off table showing the number of prizes versus the value of each prize.

Given:

Amount of door prize fund = $170
Possible numbers of prizes: 97, 20, 10, 5, 4, 3, 2, 1

Find:

Table of trade-offs of number of prizes versus value of each prize.

Solution:

The trade-off is that the more door prizes they award, the smaller each one will be.

If they give each person a door prize and have 97 of them

$$\$170 \div 97 = \$1.75$$

If they award 20 prizes

$$\$170 \div 20 = \$8.50$$

If they give 10 prizes

$$\$170 \div 10 = \$17.00$$

For 5 prizes

$$\$170 \div 5 = \$34.00$$

For 4 prizes

$170 ÷ 4 = $42.50

For 3 prizes

$170 ÷ 3 = $56.67

For 2 prizes

$170 ÷ 2 = $85.00

For 1 prize

$170 ÷ 1 = $170

Answer:

The trade-off table is as follows:

Number of Prizes	*Value of Each Prize*
97	*$ 1.75*
20	*8.50*
10	*17.00*
5	*34.00*
4	*42.50*
3	*56.67*
2	*85.00*
1	*170.00*

The choice from among these possible ways is up to Edith and Felicia. In choosing the number of prizes they will be guided by what they think people expect. If they think everyone expects to get a prize, they will award 97 of them at $1.75 each. If they sense that people think that $1.75 prizes are just cheap trinkets, they will award fewer but more expensive prizes.

A lot of things in life involve trade-offs. Often the trade-offs aren't as obvious as those in packing a suitcase or dividing door-prize money. For instance, there is the question of energy versus the environment. If we try to get energy very cheaply, such as by burning coal without scrubbing systems, then the environment gets very polluted. On the other hand, if we have very tight pollution standards, then the environment stays clean, but the coal-burning process becomes very expensive because of all the sophisticated antipollution devices.

Closer to home, there are many trade-off problems that come up in family budgeting. One problem that occurs very often is the trade-off between spending money or saving it. The more you spend now, the longer you're going to have to wait for the things you're saving for.

For example, the Robinson family is trying to figure out their budget. They are trying to save $9000 for a down payment on a house, but they also have to meet the expenses of their day-to-day living.

Jim Robinson is a machinist and his take-home pay is $1078.70 per month after taxes and union dues. Right off the top, he has to write a rent check for $220 each month for the family's apartment. His wife, Angela, has kept track of the grocery expenses and has found that they average $310 per month. Clothing for themselves and for their two children, Mark and Christine, costs a total of $1680 per year.

They also have expenses for their family car, which Jim drives to work every day and Angela uses on the weekends to go shopping. Their Gazelle 6 auto is 3 years old now, and the payments are all completed, so Jim only has to worry about operating expenses. He spends $35 per month on gas and $980 per year on repairs. His yearly bills for car insurance are $290 and for licenses $65. Parking costs him an average of $22 a month.

Jim had looked at a lot of life insurance plans and decided that term insurance was the best deal. He pays a premium of $190 per year. Jim also has health insurance through his employer, Consolidated Widget Manufacturing, which he pays $55 a month for.

Jim and Angela consider all the items we have mentioned so far as necessities. Let's total them up on a monthly basis. We'll divide all the expenses that are annual ones by 12 to make them monthly.

Rent	$220.00
Food	310.00
Clothing ($1680 ÷ 12)	140.00
Gas for car	35.00
Car repairs (980 ÷ 12)	81.67
Car insurance (290 ÷ 12)	24.17
Car licenses ($65 ÷ 12)	5.42
Parking	22.00
Term life insurance ($190 ÷ 12)	15.83
Health insurance	55.00
Total	$909.09

Besides the above expenses the Robinsons find that little miscellaneous necessities come up every year. There are doctor and dentist bills not covered by the health insurance, tools Jim has to buy for work, Angela's beauty parlor appointments, and school supplies for Mark and Christine. These little things totaled up to $328 last year, and the Robinsons think they will be about the same this year. On a monthly basis these miscellaneous expenses are $328 ÷ 12 = $27.33. Adding this to the expenses already totaled, the Robinsons have

$$\$909.09 + 27.33 = \$936.42$$

This is their total expense for necessities for the average month.

The money that the Robinsons have left over after necessities each month is

$$\$1078.70 - 936.42 = \$142.28$$

This $142.28 is money the Robinsons can save toward buying a house, or they can spend it on recreation. If they spend some of it on recreation, they will have to wait longer to buy the house. That's the trade-off—they have to trade enjoying the money now by spending it on recreation for enjoying the house sooner.

Suppose they don't spend any money at all on recreation, and they save $142.28 every month. Then at the end of a year they have

$$\$142.28 \times 12 = \$1707.36$$

If they put this money into a certificate of deposit each year at 8% interest, it will grow from compound interest. We can figure out how much they'll have at the end of 5 years by using the 8% interest table at the back of the book. For $n = 5$ we find that $Y = 5.867$. The future worth at the end of 5 years is

$$\begin{aligned} F &= AY \\ &= \$1707.36 \times 5.867 \\ &= \$10{,}017.08 \end{aligned}$$

They would then have enough for a down payment with $1017.08 left over. Saving all the money lets them buy the house in 5 years.

If instead the Robinsons decided to take a vacation every year for $1500, they would have to subtract this from the money they were saving for the house.

$$\$1707.36 - 1500.00 = \$207.36$$

How many years would it take them to save $9000 if they saved this $207.36 every year? To find out, let's calculate the Y factor and look in the 8% table again.

$$\begin{aligned} Y &= F \div A \\ &= \$9000 \div \$207.36 \\ &= 43.403 \end{aligned}$$

Looking for this Y factor in the 8% table, we see that it is a little less than $n = 20$. Thus the Robinsons would have to wait 20 years to buy the house if they took a $1500 vacation every year. If they do, they better buy a small house. Mark and Christine will be grown up and gone by then.

Let's look at some cases between these extremes.

If the Robinsons spend $300 for recreation each year, there will be $1407.36 left to save for the house. The Y factor is

$$\begin{aligned} Y &= \$9000 \div \$1407.36 \\ &= 6.395 \end{aligned}$$

Looking in the table again, we find n = 6 years.

Or if they spend $700 for recreation, there is $1007.36 left to save and

$$\begin{aligned} Y &= \$9000 \div \$1007.36 \\ &= 8.934 \end{aligned}$$

Looking in the table, we see it will take them 7 years to save the money.

Similarly, if they spend $1100 for recreation, there is $607.36 left to save and

$$\begin{aligned} Y &= \$9000 \div \$607.36 \\ &= 14.818 \end{aligned}$$

From the table we see it will take them 10 years to save the money.

We can summarize these numbers we've calculated in two columns as follows:

Dollars Spent for Recreation	Years Waited to Buy House
0	5
300	6
700	7
1100	10
1500	20

This two-column table clearly shows us the trade-offs. The numbers tell us how many years longer the Robinsons have to wait for the house for each additional amount they spend on recreation. This is what is known as the "terms of the trade."

The Robinsons went through these calculations and looked at this table of numbers. They wanted their own home as soon as possible, but they didn't want to have 5 years with no recreation at all—not ever being able to go to movies, go out to eat, or take Mark and Christine to the amusement park. They decided to limit themselves to $300 per year for recreation and buy the house in 6 years. They decided that getting the house only 1 year sooner through having no recreation at all was too big a sacrifice.

They did have to do without going away on vacation but they decided it was worth it. The key to their decision was figuring out the numbers involved in the trade-offs. They were then able to make an intelligent decision, knowing the terms of trade.

Now that you've seen how trade-offs are analyzed, let's try a couple problems.

Problem 48

The Watson family is looking at their budget for the year, and they find that after accounting for necessities, there's $895 left. They are trying to figure out the trade-offs of spending money for entertainment versus buying a car sooner. The used car they want costs $3800, and they want to pay cash so they can avoid the interest payments. The bank where they save pays 5.5% interest compounded annually. How many years will they have to wait for the car if they spend $100 per year for entertainment? How about $300? $600? What if they save it all?

Given:

Used automobile costs $3800
Bank pays 5.5% interest, compounded annually
Income left after necessities = $895
Possible expenditures for entertainment are 0, $100, $300, and $600

Find:

Years they must wait to buy the car for each level of entertainment expense.

Solution:

This is similar to the case of the Robinsons, which we just discussed. The problem calls for using the Y column of the 5.5% interest table.

In the case where all $895 is saved, we have

$$\begin{aligned} Y &= F \div A \\ &= \$3800 \div \$895 \\ &= 4.246 \end{aligned}$$

Looking this up in the table, we find that n = 4 years.

When $100 is spent on entertainment, there is $795 left, and we have

$$Y = \$3800 \div \$795$$
$$= 4.800$$

From the table we have n = 5 years.

When $300 is spent on entertainment, there is $595 left.

$$Y = \$3800 \div \$595$$
$$= 6.387$$

The table gives n = 6 years.

When $600 is spent on entertainment, there is $295 left.

$$Y = \$3800 \div \$295$$
$$= 12.881$$

For this value of Y, the table shows n = 10 years.

Answer:

The trade-offs are summarized in the table below.

Dollars Spent for Entertainment	*Years Waited to Buy Car*
0	*4*
100	*5*
300	*6*
600	*10*

As in the case of the Robinsons, the decision on how much to spend on entertainment will depend on how badly they want to reach their savings objective. People are better prepared to make this decision knowing the trade-offs.

Problem 49

Tom Arnold is thinking of buying a boat and he wants to pay cash. The price of the boat is $7300.

Tom has $840 left each quarter, which he can spend on entertainment or save for his boat. The bank where he saves the money pays 6% compounded quarterly. When will he get the boat if he saves all of his money? If he spends $100, $300, $600?

Given:

Boat costs $7300
Bank pays 6% interest compounded quarterly
Income left after necessities = $840 per quarter
Possible expenditures for entertainment are 0, $100, $300, $600 per quarter

Find:

Years he must wait to buy the boat for each level of entertainment expense.

Solution:

This is similar to the previous problem except that the interest is compounded quarterly. The quarterly interest rate is

$$6\% \div 4 = 1.5\%$$

Therefore, we will be using the Y column of the 1.5% table in the back of the book.

For the case where he saves the whole $840

$$\begin{aligned} Y &= F \div A \\ &= \$7300 \div \$840 \\ &= 8.690 \end{aligned}$$

The table shows that n = 9 quarters; he waits 2.25 years.

If he spends $100, he saves $740 and

$$\begin{aligned} Y &= \$7300 \div \$740 \\ &= 9.865 \end{aligned}$$

Checking this in the table, we find n = 10 quarters. This is 2.5 years.

If he spends $300 he saves $540 and

$$\begin{aligned} Y &= \$7300 \div \$540 \\ &= 13.519 \end{aligned}$$

From the table we see n = 12 quarters. This is 3 years.

If he spends $600 on entertainment, then he only saves $240.

$$Y = \$7300 \div \$240$$
$$= 30.417$$

The table shows that n = 26 quarters. This equals 6.5 years.

Answer:

The trade-offs are summarized below.

Dollars Spent for Entertainment	*Years Waited to Buy Boat*
0	*2.25*
100	*2.50*
300	*3.00*
600	*6.50*

As in the other problems, we found that the less Tom Arnold saved, the longer it took him to get together enough money for his boat. This was another typical trade-off situation.

In our look at trade-offs we have concentrated on cases where there are two factors that are traded off against each other. But what do you do when there are more than two factors? How do you choose from among the trade-off possibilities? The next section looks at these questions.

Decision making

Up to now we have talked mainly about how to lay out the sets of choices when you have a trade-off. But how do you decide?

You have to decide among the choices on the basis of what you're trying to do—your objective. You may often have more than one objective.

For instance, in the example about the Robinson family's budget, they had two objectives. Their most important one was to save money to buy a house. But they also wanted to have at least a little entertainment. This led them to the choice of spending $300 per year on entertainment and saving the rest.

Weighting and Scoring

For more complicated choices, where there are more than two objectives, it usually helps to make the choices using numbers. The way this is done is with "weighting and scoring systems."

To show you how weighting and scoring systems work, let's consider an example. Suppose Don Gordon is buying a new car and he wants one that has a low price, looks nice, and gives good fuel economy. These are his three objectives. The next thing for him to do is figure out how important each of these objectives is to him, on a scale of 1 to 10.

Price is the most important to him so he gives it a weight of 10. Looks are not quite so important so he gives that a 6. Fuel economy isn't really too important to him so he gives that a weighting of 2. These are the weightings of the objectives he will use.

Don looks at three cars. The Calamari Spyder is an expensive foreign sports car; the Frugal-4 is a small economy-priced car; and the Provincial-6 is a middle-sized car that has been designed for better fuel economy. In looking at each of these cars, Don scores them 1 to 10 for price, looks, and fuel economy.

The Calamari Spyder costs $19,000, making it by far the most expensive of the three cars. Don scores it 2 out of a possible 10 in the price category. However, it does look sexier than any car Don has ever seen, so he gives it a 10 for looks. Its fuel economy is a little worse than the average car, so he scores that a 4. Now he totals up the rating for the Calamari Spyder. Its rating for price is the weighting for price, which is 10, times the score for price, which is 2. The rating is then

$$10 \times 2 = 20$$

For the Calamari Spyder's looks the weighting is 6 and the score is 10, so the rating is

$$6 \times 10 = 60$$

For the Calamari Spyder's fuel economy the weighting is 2 and the score is 4, so the rating is

$$2 \times 4 = 8$$

Adding these together to get the Calamari Spyder's total rating, Don gets

Price	20
Looks	60
Fuel economy	8
Total rating	88

Next Don looks at the Frugal-4. He can get one for $5700, making it the lowest priced of the three. He gives it a score of 10 for price. It really isn't a very pretty car at all, so he scores it 2 for looks. It's close to the top of all cars for fuel economy, so he scores it a 9. Totaling up the rating for the Frugal-4, Don gets

Price 10 × 10	= 100
Looks 6 × 2	= 12
Fuel economy 2 × 9	= 18
Total rating	130

Finally, Don scores the Provincial-6. It costs $9300, intermediate between the other two cars, and he scores it a 6 for price. It has average looks so he gives it a 5. Its fuel economy is almost as good as the Frugal-4, so he gives it an 8. Totaling its rating gives

Price 10 × 6	= 60
Looks 6 × 5	= 30
Fuel economy 2 × 8	= 16
Total rating	106

Summarizing the total ratings, Don finds

Calamari Spyder	88
Frugal-4	130
Provincial-6	106

Since the Frugal-4 has the highest total rating, he buys it. He realizes that the other two cars were prettier, but he knows price is more important to him. After all, that's the way he set up the weighting numbers. It would have been nice for Don to have a better looking car, but he couldn't get one at a good price. Those are the trade-offs.

Using the weighting and scoring system helped Don Gordon make a tough choice. Since he was the one who picked the weightings and scorings, he knew it was the right decision.

Let's try some problems in weightings and scorings, so you get the hang of using them.

Problem 50

Herman and Emily Peterson are buying a new house. They are using a weighting and scoring system to pick from among their three choices. They decide to assign a weighting of 10 for location, 9 for price, and 6 for condition.

One house is a colonial style that sells for $75,000. The house is in good condition and centrally located, but the neighborhood is run-down. They score it 4 for location, 9 for price, and 9 for condition.

The next house they look at is a tract house in a new development. The development is a nice neighborhood, but it's far out in the suburbs, making commuting difficult. They score its location 4. It sells for $80,000 and they score this price as 7. It's almost brand-new but is poorly constructed. They rate its condition as 8.

Finally, they look at a reconditioned house in a good neighborhood, which is centrally located. They score its location 9. Its price is $90,000; they score this 5. Even though it has been reconditioned, it still needs some work. They rate its condition as 4.

What are the overall ratings of the three houses? Which one do they buy?

Given:

Weightings:

Location	10
Price	9
Condition	6

Scorings:

Colonial house—location 4, price 9, condition 9
Tract house—location 4, price 7, condition 8
Reconditioned house—location 9, price 5, condition 4

Find:

Overall ratings of the three houses.
Choice of house.

Solution:

Set up the weighting and scoring as we did in the previous discussion of Don Gordon's car.

Colonial house

Location 10×4	= 40
Price 9×9	= 81
Condition 6×9	= 54
Total rating	175

Tract house

Location 10×4	= 40
Price 9×7	= 63
Condition 6×8	= 48
Total rating	151

Reconditioned house

Location 10×9	= 90
Price 9×5	= 45
Condition 6×4	= 24
Total rating	159

Answer:

The total ratings are:

Colonial house	*175*
Tract house	*151*
Reconditioned house	*159*

The Petersons buy the colonial house.

Problem 51

Julie Greenberg is choosing a new job. She is using a weighting and scoring system to choose between two possible employers. She considers the most important thing to be the type of work—whether it's something she likes to do. She weights the type of work 10. Next in importance is the chance for advancement. She weights this as 9. Salary is not quite so important to her, and she weights it as 8. She also considers how close the office is to her home, but she weights this as only 2.

One employer is the *Daily Bugle*, the local newspaper. Julie is interested in working on a newspaper, and her degree is in journalism. She scores the type of work as 10. Unfortunately, if she joins the *Daily Bugle*, she will have to start from the bottom as a copy girl. Most people move up from there to being a cub reporter. She scores the advancement possibilities as 8, but the starting salary is very low, and she scores it as 2. The *Daily Bugle* is near her home so she scores the commuting distance as 10.

The other job she has been offered is as a technical writer for Consolidated Electronics. She would be writing handbooks for the various electronic products that the company makes. The type of work is interesting to her since it involves writing, but not as interesting as newspaper work. She scores it a 6. The advancement opportunities are only fair at Consolidated Electronics. People who begin as technical writers tend to stay technical writers. She scores advancement as 6. The starting salary at Consolidated Electronics is very high, and she rates it as 10. The office is a moderate distance from home, and she scores commuting distance as 5.

What are the overall ratings of the two jobs? Which one does Julie choose?

Given:

Weightings:

Type	10
Advancement	9

Salary	8
Commuting distance	2

Scorings:

Daily Bugle—Type 10, advancement 8, salary 2, commuting distance 10

Consolidated Electronics—Type 6, advancement 6, salary 10, commuting distance 5

Find:

Overall rating of each employer.
Choice of employer.

Solution:

This problem is set up like the previous one. It's just a question of multiplying weightings by scores.

Daily Bugle

Type 10 × 10	= 100
Advancement 9 × 8	= 72
Salary 8 × 2	= 16
Commuting distance 2 × 10	= 20
Total rating	208

Consolidated Electronics

Type 10 × 6	= 60
Advancement 9 × 6	= 54
Salary 8 × 10	= 80
Commuting distance 2 × 5	= 10
Total rating	= 204

Answer:

The total ratings for the two employers are:

Daily Bugle 208

Consolidated Electronics 204

Julie chooses the Daily Bugle.

This problem resulted in a very close decision, but Julie stuck by her weighting and scoring system and went with the *Daily Bugle*. It is lucky she had a weighting and scoring system to help her decide. Otherwise, she might have been tempted to postpone this difficult decision. Putting off the decision might have meant days without pay and sleepless nights while deciding. She avoided this by making up her mind.

Problem 52

Andrea Pollack is shopping for a new cocktail dress. After looking at several stores, she narrows the choice down to two. She decides to use a weighting and scoring system to pick one of the two. Her first concern is the style, which she weights as 10. Next she is thinking of price, which she weights as 8. The fit is not quite so important to her since she can have the dress altered if necessary; she weights it 5. She is also somewhat concerned with the color and weights it 4.

The green dress she has found is just the style she was looking for, and she scores it 10. It is also more expensive, and she scores it 2 on price. The dress fits extremely well and would not need alterations, so she scores that 10; but green is not one of her favorite colors so she scores it 3.

The red dress appears to be last year's style, but it looks okay, so she scores it 3 for style. It's a real bargain, so she scores it 10 for price. It fits pretty well, so she scores it 8, and since red is her favorite color, she scores it 10.

What are the total ratings for the two dresses? Which one does she choose?

Given:

Weightings:

Style	10
Price	8
Fit	5
Color	4

Scorings:

Green dress—style 10, price 2, fit 10, color 3
Red dress—style 3, price 10, fit 8, color 10

Find:

Total ratings for the two dresses.
Choice of dress.

Solution:

The set up is the same as the previous weighting and scoring problems.

Green dress

Style 10 × 10	= 100
Price 8 × 2	= 16
Fit 5 × 10	= 50
Color 4 × 3	= 12
Total rating	178

Red dress

Style 10 × 3	= 30
Price 8 × 10	= 80
Fit 5 × 8	= 40
Color 4 × 10	= 40
Total rating	= 190

Answer:

The total ratings are:

Green dress	*178*
Red dress	*190*

Andrea buys the red dress.

Problem 53

Now that their children have grown, Harry and Samantha Stone want to sell their present house and buy a smaller one closer to the center of the city. They have found four houses, all in the same neighborhood. Their decision will be based on price and condition of the house. They are using a weighting and scoring system, and they weight price as 10 and condition as 6.

The first house they are considering is a frame house. It costs only $75,000, but it's in poor condition. They score it 10 on price and 1 on condition.

The second one is a reconditioned brownstone. It costs $87,000 and is in moderately good shape. They score it 6 on price and 6 on condition.

The third house is a relatively new brick house, which costs $98,000. It is in excellent condition. They score it 3 on price and 10 on condition.

The fourth house is an old stucco house, which recently has been completely reconditioned. It is selling for $82,000. They score its price as 8 and its condition as 7.

What are the total ratings of the four houses? Which one do they buy?

Given:

Weightings:

Price	10
Condition	6

Scorings:

Frame house—price 10, condition 1
Brownstone—price 6, condition 6
Brick house—price 3, condition 10
Stucco house—price 8, condition 7

Find:

Total ratings of the four houses.
Choice of house.

Solution:

The set up is the same as the earlier problems. The only difference is that there are four things to be rated.

Frame house	
Price 10×10	= 100
Condition 6×1	= 6
Total rating	106

Brownstone	
Price 10×6	= 60
Condition 6×6	= 36
Total rating	96

Brick house

Price 10 × 3	=	30
Condition 6 × 10	=	60
Total rating		90

Stucco house

Price 10 × 8	=	80
Condition 6 × 7	=	42
Total rating		122

Answer:

The total ratings are:

Frame house	*106*
Brownstone	*96*
Brick house	*90*
Stucco house	*122*

The Stones buy the stucco house.

Having worked these weighting and scoring problems, you are now in a position to do weightings and scorings of your own. You'll find that they'll make it easier to make tough decisions—the kind that people worry over for too long without making up their minds. The success you have in using weighting and scoring systems will depend on how carefully you choose your rating factors, what weights you give them, and what scores you give the choices. If they are your best judgments, then the total rating should be the decision you want.

Trade-offs are no problem if you recognize them and learn to choose among them, as we did in this chapter.

8. SUMMING UP

What Do You Have to Gain?

Now that you have put in the work to read this book and work the sample problems, you're probably wondering what comes next. Well, that is really up to you. You can go one of two ways:

1. *You can apply what you have learned.* You can start thinking with numbers. When you have an interest problem or a percentage problem, you can take out your calculator and figure it out. If you're making a decision, you can calculate the mathematical expectations or you can set up a weighting and scoring system. You now know the methods, and life supplies the problems. With your calculator in your hand, you can get the answers.

2. *You can forget what you learned if you don't use it.* If you don't apply what you learned in this book,

you're going to forget it. If you make decisions by whim or default and you never stop to calculate where you stand, then this book will not have done you any good.

In making your choice between these two ways, you must ask yourself, "What do I have to gain by thinking with numbers?"

The first thing you can be assured of is that when you think with numbers you are going to wind up with more money. You'll stop being ripped off by installment payments, gambling games, and bad deals on insurance. You'll know what kind of return to expect from your investments. You'll be able to make the right decision when you're thinking of changing jobs.

Thinking with numbers will also make you more self-confident. You'll find that you are figuring things out for yourself, rather than taking someone else's word for it. You'll also find you are right more of the time. There's nothing that improves your self-assurance more than seeing your plans work out right.

Some People Never Learn

Even if you become very successful from thinking with numbers, you're going to find some people who don't like it. First of all, there will be those people who want to rip you off. If they aren't able to rip you off any more because you can now think with numbers, they won't be too happy about it.

You'll also find a lot of "sour grapes" people out there, who can't think with numbers themselves and are jealous because you can. They will tell you that you've become too hard and calculating and that you should instead make decisions "from your heart." They'll tell you that making calculations takes too much time and effort and they would rather "play it by ear." Well, they are dead wrong!

When you hear people talking that way, just smile and walk away. Then you can laugh all the way to the bank.

A Better World for Us All

What do you think would happen if everyone thought with numbers? We'd be a lot better off—that's what would happen. Can you imagine a world where merchants watch the percentages on their profits because they know everyone else can calculate those percentages too; a world where gambling joints close down because nobody wants to make a bet with a negative mathematical expectation; a world where life insurance salesmen don't bug people because everyone buys only the insurance he needs; a world where personal loan companies go broke because people would rather save their money and wait for what they want than pay 30% interest rates; a world where real estate salesmen have to watch what they say because everyone knows how to calculate mortgage payments; a world where banks pay higher interest to depositers than they do now and demand less interest of borrowers, because everyone knows how to calculate interest and will shop around for the best rates; a world where stockbrokers are less eager to push "hot deals" because everyone knows how to calculate growth rates; a world where politicians watch their promises and think twice before raising taxes because people know the difference between first costs and operating costs—in short, a world with more mutual respect?

If you start thinking with numbers and tell your friends about it too, we're on our way. Take your calculator in hand and go get 'em!

THE END

APPENDIX
COMPOUND INTEREST TABLES

Definitions and Formulas for Using the Tables

Definitions

P = Principal, an amount of money here in the present

n = Number of periods. (The periods can be years, months, or quarters, but your interest rates have to be quoted as yearly, monthly, or quarterly to go with them.)

F = Future worth, at the end of n periods

A = Annuity, the amount of money periodically put in or taken out

Formulas

To find future worth F from principal P

$$F = PU$$

where U is the factor tabulated in the second column.

To find principal P from future worth F

$$P = FV$$

where V is the factor tabulated in the third column.

To find annuity A from future worth F

$$A = FW$$

where W is the factor tabulated in the fourth column.

To find annuity A from principal P

$$A = PX$$

where X is the factor tabulated in the fifth column.

To find future worth F from annuity A

$$F = AY$$

where Y is the factor tabulated in the sixth column.

To find principal P from annuity A

$$P = AZ$$

where Z is the factor tabulated in the seventh column.

1% COMPOUND INTEREST TABLE

n	U	V	W	X	Y	Z	n
1	1.010	0.9901	1.00000	1.01000	1.000	0.990	1
2	1.020	0.9803	0.49751	0.50751	2.010	1.970	2
3	1.030	0.9706	0.33002	0.34002	3.030	2.941	3
4	1.041	0.9610	0.24628	0.25628	4.060	3.902	4
5	1.051	0.9515	0.19604	0.20604	5.101	4.853	5
6	1.062	0.9420	0.16255	0.17255	6.152	5.795	6
7	1.072	0.9327	0.13863	0.14863	7.214	6.728	7
8	1.083	0.9235	0.12069	0.13069	8.286	7.652	8
9	1.094	0.9143	0.10674	0.11674	9.369	8.566	9
10	1.105	0.9053	0.09558	0.10558	10.462	9.471	10
11	1.116	0.8963	0.08645	0.09645	11.567	10.368	11
12	1.127	0.8874	0.07885	0.08885	12.683	11.255	12
13	1.138	0.8787	0.07241	0.08241	13.809	12.134	13
14	1.149	0.8700	0.06690	0.07690	14.947	13.004	14
15	1.161	0.8613	0.06212	0.07212	16.097	13.865	15
16	1.173	0.8528	0.05794	0.06794	17.258	14.718	16
17	1.184	0.8444	0.05426	0.06426	18.430	15.562	17
18	1.196	0.8360	0.05098	0.06098	19.615	16.398	18
19	1.208	0.8277	0.04805	0.05805	20.811	17.226	19
20	1.220	0.8195	0.04542	0.05542	22.019	18.046	20
21	1.232	0.8114	0.04303	0.05303	23.239	18.857	21
22	1.245	0.8034	0.04086	0.05086	24.472	19.660	22
23	1.257	0.7954	0.03889	0.04889	25.716	20.456	23
24	1.270	0.7876	0.03707	0.04707	26.973	21.243	24
25	1.282	0.7798	0.03541	0.04541	28.243	22.023	25
26	1.295	0.7720	0.03387	0.04387	29.526	22.795	26
27	1.308	0.7644	0.03245	0.04245	30.821	23.560	27
28	1.321	0.7568	0.03112	0.04112	32.129	24.316	28
29	1.335	0.7493	0.02990	0.03990	33.450	25.066	29
30	1.348	0.7419	0.02875	0.03875	34.785	25.808	30
31	1.361	0.7346	0.02768	0.03768	36.133	26.542	31
32	1.375	0.7273	0.02667	0.03667	37.494	27.270	32
33	1.389	0.7201	0.02573	0.03573	38.869	27.990	33
34	1.403	0.7130	0.02484	0.03484	40.258	28.703	34
35	1.417	0.7059	0.02400	0.03400	41.660	29.409	35
40	1.489	0.6717	0.02046	0.03046	48.886	32.835	40
45	1.565	0.6391	0.01771	0.02771	56.481	36.095	45
50	1.645	0.6080	0.01551	0.02551	64.463	39.196	50
55	1.729	0.5785	0.01373	0.02373	72.852	42.147	55
60	1.817	0.5504	0.01224	0.02224	81.670	44.955	60
65	1.909	0.5237	0.01100	0.02100	90.937	47.627	65
70	2.007	0.4983	0.00993	0.01993	100.676	50.169	70
75	2.109	0.4741	0.00902	0.01902	110.913	52.587	75
80	2.217	0.4511	0.00822	0.01822	121.672	54.888	80
85	2.330	0.4292	0.00752	0.01752	132.979	57.078	85
90	2.449	0.4084	0.00690	0.01690	144.863	59.161	90
95	2.574	0.3886	0.00636	0.01636	157.354	61.143	95
100	2.705	0.3697	0.00587	0.01587	170.481	63.029	100

1.25% COMPOUND INTEREST TABLE

n	U	V	W	X	Y	Z	n
1	1.012	0.9877	1.00000	1.01250	1.000	0.988	**1**
2	1.025	0.9755	0.49689	0.50939	2.012	1.963	**2**
3	1.038	0.9634	0.32920	0.34170	3.038	2.927	**3**
4	1.051	0.9515	0.24536	0.25786	4.076	3.878	**4**
5	1.064	0.9398	0.19506	0.20756	5.127	4.818	**5**
6	1.077	0.9282	0.16153	0.17403	6.191	5.746	**6**
7	1.091	0.9167	0.13759	0.15009	7.268	6.663	**7**
8	1.104	0.9054	0.11963	0.13213	8.359	7.568	**8**
9	1.118	0.8942	0.10567	0.11817	9.463	8.462	**9**
10	1.132	0.8832	0.09450	0.10700	10.582	9.346	**10**
11	1.146	0.8723	0.08537	0.09787	11.714	10.218	**11**
12	1.161	0.8615	0.07776	0.09026	12.860	11.079	**12**
13	1.175	0.8509	0.07132	0.08382	14.021	11.930	**13**
14	1.190	0.8404	0.06581	0.07831	15.196	12.771	**14**
15	1.205	0.8300	0.06103	0.07353	16.386	13.601	**15**
16	1.220	0.8197	0.05685	0.06935	17.591	14.420	**16**
17	1.235	0.8096	0.05316	0.06566	18.811	15.230	**17**
18	1.251	0.7996	0.04988	0.06238	20.046	16.030	**18**
19	1.266	0.7898	0.04696	0.05946	21.297	16.819	**19**
20	1.282	0.7800	0.04432	0.05682	22.563	17.599	**20**
21	1.298	0.7704	0.04194	0.05444	23.845	18.370	**21**
22	1.314	0.7609	0.03977	0.05227	25.143	19.131	**22**
23	1.331	0.7515	0.03780	0.05030	26.457	19.882	**23**
24	1.347	0.7422	0.03599	0.04849	27.788	20.624	**24**
25	1.364	0.7330	0.03432	0.04682	29.135	21.357	**25**
26	1.381	0.7240	0.03279	0.04529	30.500	22.081	**26**
27	1.399	0.7150	0.03137	0.04387	31.881	22.796	**27**
28	1.416	0.7062	0.03005	0.04255	33.279	23.503	**28**
29	1.434	0.6975	0.02882	0.04132	34.695	24.200	**29**
30	1.452	0.6889	0.02768	0.04018	36.129	24.889	**30**
31	1.470	0.6804	0.02661	0.03911	37.581	25.569	**31**
32	1.488	0.6720	0.02561	0.03811	39.050	26.241	**32**
33	1.507	0.6637	0.02467	0.03717	40.539	26.905	**33**
34	1.526	0.6555	0.02378	0.03628	42.045	27.560	**34**
35	1.545	0.6474	0.02295	0.03545	43.571	28.208	**35**
40	1.644	0.6084	0.01942	0.03192	51.490	31.327	**40**
45	1.749	0.5718	0.01669	0.02919	59.916	34.258	**45**
50	1.861	0.5373	0.01452	0.02702	68.882	37.013	**50**
55	1.980	0.5050	0.01275	0.02525	78.422	39.602	**55**
60	2.107	0.4746	0.01129	0.02379	88.575	42.035	**60**
65	2.242	0.4460	0.01006	0.02256	99.377	44.321	**65**
70	2.386	0.4191	0.00902	0.02152	110.872	46.470	**70**
75	2.539	0.3939	0.00812	0.02062	123.103	48.489	**75**
80	2.701	0.3702	0.00735	0.01985	136.119	50.387	**80**
85	2.875	0.3479	0.00667	0.01917	149.968	52.170	**85**
90	3.059	0.3269	0.00607	0.01857	164.705	53.846	**90**
95	3.255	0.3072	0.00554	0.01804	180.386	55.421	**95**
100	3.463	0.2887	0.00507	0.01757	197.072	56.901	**100**

1.5% COMPOUND INTEREST TABLE

n	U	V	W	X	Y	Z	n
1	1.015	0.9852	1.00000	1.01500	1.000	0.985	**1**
2	1.030	0.9707	0.49628	0.51128	2.015	1.956	**2**
3	1.046	0.9563	0.32838	0.34338	3.045	2.912	**3**
4	1.061	0.9422	0.24444	0.25944	4.091	3.854	**4**
5	1.077	0.9283	0.19409	0.20909	5.152	4.783	**5**
6	1.093	0.9145	0.16053	0.17553	6.230	5.697	**6**
7	1.110	0.9010	0.13656	0.15156	7.323	6.598	**7**
8	1.126	0.8877	0.11858	0.13358	8.433	7.486	**8**
9	1.143	0.8746	0.10461	0.11961	9.559	8.361	**9**
10	1.161	0.8617	0.09343	0.10843	10.703	9.222	**10**
11	1.178	0.8489	0.08429	0.09929	11.863	10.071	**11**
12	1.196	0.8364	0.07668	0.09168	13.041	10.908	**12**
13	1.214	0.8240	0.07024	0.08524	14.237	11.732	**13**
14	1.232	0.8118	0.06472	0.07972	15.450	12.543	**14**
15	1.250	0.7999	0.05994	0.07494	16.682	13.343	**15**
16	1.269	0.7880	0.05577	0.07077	17.932	14.131	**16**
17	1.288	0.7764	0.05208	0.06708	19.201	14.908	**17**
18	1.307	0.7649	0.04881	0.06381	20.489	15.673	**18**
19	1.327	0.7536	0.04588	0.06088	21.797	16.426	**19**
20	1.347	0.7425	0.04325	0.05825	23.124	17.169	**20**
21	1.367	0.7315	0.04087	0.05587	24.471	17.900	**21**
22	1.388	0.7207	0.03870	0.05370	25.838	18.621	**22**
23	1.408	0.7100	0.03673	0.05173	27.225	19.331	**23**
24	1.430	0.6995	0.03492	0.04992	28.634	20.030	**24**
25	1.451	0.6892	0.03326	0.04826	30.063	20.720	**25**
26	1.473	0.6790	0.03173	0.04673	31.514	21.399	**26**
27	1.495	0.6690	0.03032	0.04532	32.987	22.068	**27**
28	1.517	0.6591	0.02900	0.04400	34.481	22.727	**28**
29	1.540	0.6494	0.02778	0.04278	35.999	23.376	**29**
30	1.563	0.6398	0.02664	0.04164	37.539	24.016	**30**
31	1.587	0.6303	0.02557	0.04057	39.102	24.646	**31**
32	1.610	0.6210	0.02458	0.03958	40.688	25.267	**32**
33	1.634	0.6118	0.02364	0.03864	42.299	25.879	**33**
34	1.659	0.6028	0.02276	0.03776	43.933	26.482	**34**
35	1.684	0.5939	0.02193	0.03693	45.592	27.076	**35**
40	1.814	0.5513	0.01843	0.03343	54.268	29.916	**40**
45	1.954	0.5117	0.01572	0.03072	63.614	32.552	**45**
50	2.105	0.4750	0.01357	0.02857	73.683	35.000	**50**
55	2.268	0.4409	0.01183	0.02683	84.530	37.271	**55**
60	2.443	0.4093	0.01039	0.02539	96.215	39.380	**60**
65	2.632	0.3799	0.00919	0.02419	108.803	41.338	**65**
70	2.835	0.3527	0.00817	0.02317	122.364	43.155	**70**
75	3.055	0.3274	0.00730	0.02230	136.973	44.842	**75**
80	3.291	0.3039	0.00655	0.02155	152.711	46.407	**80**
85	3.545	0.2821	0.00589	0.02089	169.665	47.861	**85**
90	3.819	0.2619	0.00532	0.02032	187.930	49.210	**90**
95	4.114	0.2431	0.00482	0.01982	207.606	50.462	**95**
100	4.432	0.2256	0.00437	0.01937	228.803	51.625	**100**

1.75% COMPOUND INTEREST TABLE

n	U	V	W	X	Y	Z	n
1	1.018	0.9828	1.00000	1.01750	1.000	0.983	1
2	1.035	0.9659	0.49566	0.51316	2.018	1.949	2
3	1.053	0.9493	0.32757	0.34507	3.053	2.898	3
4	1.072	0.9330	0.24353	0.26103	4.106	3.831	4
5	1.091	0.9169	0.19312	0.21062	5.178	4.748	5
6	1.110	0.9011	0.15952	0.17702	6.269	5.649	6
7	1.129	0.8856	0.13553	0.15303	7.378	6.535	7
8	1.149	0.8704	0.11754	0.13504	8.508	7.405	8
9	1.169	0.8554	0.10356	0.12106	9.656	8.260	9
10	1.189	0.8407	0.09238	0.10988	10.825	9.101	10
11	1.210	0.8263	0.08323	0.10073	12.015	9.927	11
12	1.231	0.8121	0.07561	0.09311	13.225	10.740	12
13	1.253	0.7981	0.06917	0.08667	14.457	11.538	13
14	1.275	0.7844	0.06366	0.08116	15.710	12.322	14
15	1.297	0.7709	0.05888	0.07638	16.984	13.093	15
16	1.320	0.7576	0.05470	0.07220	18.282	13.850	16
17	1.343	0.7446	0.05102	0.06852	19.602	14.595	17
18	1.367	0.7318	0.04774	0.06524	20.945	15.327	18
19	1.390	0.7192	0.04482	0.06232	22.311	16.046	19
20	1.415	0.7068	0.04219	0.05969	23.702	16.753	20
21	1.440	0.6947	0.03981	0.05731	25.116	17.448	21
22	1.465	0.6827	0.03766	0.05516	26.556	18.130	22
23	1.490	0.6710	0.03569	0.05319	28.021	18.801	23
24	1.516	0.6594	0.03389	0.05139	29.511	19.461	24
25	1.543	0.6481	0.03223	0.04973	31.027	20.109	25
26	1.570	0.6369	0.03070	0.04820	32.570	20.746	26
27	1.597	0.6260	0.02929	0.04679	34.140	21.372	27
28	1.625	0.6152	0.02798	0.04548	35.738	21.987	28
29	1.654	0.6046	0.02676	0.04426	37.363	22.592	29
30	1.683	0.5942	0.02563	0.04313	39.017	23.186	30
31	1.712	0.5840	0.02457	0.04207	40.700	23.770	31
32	1.742	0.5740	0.02358	0.04108	42.412	24.344	32
33	1.773	0.5641	0.02265	0.04015	44.154	24.908	33
34	1.804	0.5544	0.02177	0.03927	45.927	25.462	34
35	1.835	0.5449	0.02095	0.03845	47.731	26.007	35
40	2.002	0.4996	0.01747	0.03497	57.234	28.594	40
45	2.183	0.4581	0.01479	0.03229	67.599	30.966	45
50	2.381	0.4200	0.01267	0.03017	78.902	33.141	50
55	2.597	0.3851	0.01096	0.02846	91.230	35.135	55
60	2.832	0.3531	0.00955	0.02705	104.675	36.964	60
65	3.088	0.3238	0.00838	0.02588	119.339	38.641	65
70	3.368	0.2969	0.00739	0.02489	135.331	40.178	70
75	3.674	0.2722	0.00655	0.02405	152.772	41.587	75
80	4.006	0.2496	0.00582	0.02332	171.794	42.880	80
85	4.369	0.2289	0.00519	0.02269	192.539	44.065	85
90	4.765	0.2098	0.00465	0.02215	215.165	45.152	90
95	5.197	0.1924	0.00417	0.02167	239.840	46.148	95
100	5.668	0.1764	0.00375	0.02125	266.752	47.061	100

2% COMPOUND INTEREST TABLE

n	U	V	W	X	Y	Z	n
1	1.020	0.9804	1.00000	1.02000	1.000	0.980	1
2	1.040	0.9612	0.49505	0.51505	2.020	1.942	2
3	1.061	0.9423	0.32675	0.34675	3.060	2.884	3
4	1.082	0.9238	0.24262	0.26262	4.122	3.808	4
5	1.104	0.9057	0.19216	0.21216	5.204	4.713	5
6	1.126	0.8880	0.15853	0.17853	6.308	5.601	6
7	1.149	0.8706	0.13451	0.15451	7.434	6.472	7
8	1.172	0.8535	0.11651	0.13651	8.583	7.325	8
9	1.195	0.8368	0.10252	0.12252	9.755	8.162	9
10	1.219	0.8203	0.09133	0.11133	10.950	8.983	10
11	1.243	0.8043	0.08218	0.10218	12.169	9.787	11
12	1.268	0.7885	0.07456	0.09456	13.412	10.575	12
13	1.294	0.7730	0.06812	0.08812	14.680	11.348	13
14	1.319	0.7579	0.06260	0.08260	15.974	12.106	14
15	1.346	0.7430	0.05783	0.07783	17.293	12.849	15
16	1.373	0.7284	0.05365	0.07365	18.639	13.578	16
17	1.400	0.7142	0.04997	0.06997	20.012	14.292	17
18	1.428	0.7002	0.04670	0.06670	21.412	14.992	18
19	1.457	0.6864	0.04378	0.06378	22.841	15.678	19
20	1.486	0.6730	0.04116	0.06116	24.297	16.351	20
21	1.516	0.6598	0.03878	0.05878	25.783	17.011	21
22	1.546	0.6468	0.03663	0.05663	27.299	17.658	22
23	1.577	0.6342	0.03467	0.05467	28.845	18.292	23
24	1.608	0.6217	0.03287	0.05287	30.422	18.914	24
25	1.641	0.6095	0.03122	0.05122	32.030	19.523	25
26	1.673	0.5976	0.02970	0.04970	33.671	20.121	26
27	1.707	0.5859	0.02829	0.04829	35.344	20.707	27
28	1.741	0.5744	0.02699	0.04699	37.051	21.281	28
29	1.776	0.5631	0.02578	0.04578	38.792	21.844	29
30	1.811	0.5521	0.02465	0.04465	40.568	22.396	30
31	1.848	0.5412	0.02360	0.04360	42.379	22.938	31
32	1.885	0.5306	0.02261	0.04261	44.227	23.468	32
33	1.922	0.5202	0.02169	0.04169	46.112	23.989	33
34	1.961	0.5100	0.02082	0.04082	48.034	24.499	34
35	2.000	0.5000	0.02000	0.04000	49.994	24.999	35
40	2.208	0.4529	0.01656	0.03656	60.402	27.355	40
45	2.438	0.4102	0.01391	0.03391	71.893	29.490	45
50	2.692	0.3715	0.01182	0.03182	84.579	31.424	50
55	2.972	0.3365	0.01014	0.03014	98.587	33.175	55
60	3.281	0.3048	0.00877	0.02877	114.052	34.761	60
65	3.623	0.2761	0.00763	0.02763	131.126	36.197	65
70	4.000	0.2500	0.00667	0.02667	149.978	37.499	70
75	4.416	0.2265	0.00586	0.02586	170.792	38.677	75
80	4.875	0.2051	0.00516	0.02516	193.772	39.745	80
85	5.383	0.1858	0.00456	0.02456	219.144	40.711	85
90	5.943	0.1683	0.00405	0.02405	247.157	41.587	90
95	6.562	0.1524	0.00360	0.02360	278.085	42.380	95
100	7.245	0.1380	0.00320	0.02320	312.232	43.098	100

2.5% COMPOUND INTEREST TABLE

n	U	V	W	X	Y	Z	n
1	1.025	0.9756	1.00000	1.02500	1.000	0.976	**1**
2	1.051	0.9518	0.49383	0.51883	2.025	1.927	**2**
3	1.077	0.9286	0.32514	0.35014	3.076	2.856	**3**
4	1.104	0.9060	0.24082	0.26582	4.153	3.762	**4**
5	1.131	0.8839	0.19025	0.21525	5.256	4.646	**5**
6	1.160	0.8623	0.15655	0.18155	6.388	5.508	**6**
7	1.189	0.8413	0.13250	0.15750	7.547	6.349	**7**
8	1.218	0.8207	0.11447	0.13947	8.736	7.170	**8**
9	1.249	0.8007	0.10046	0.12546	9.955	7.971	**9**
10	1.280	0.7812	0.08926	0.11426	11.203	8.752	**10**
11	1.312	0.7621	0.08011	0.10511	12.483	9.514	**11**
12	1.345	0.7436	0.07249	0.09749	13.796	10.258	**12**
13	1.379	0.7254	0.06605	0.09105	15.140	10.983	**13**
14	1.413	0.7077	0.06054	0.08554	16.519	11.691	**14**
15	1.448	0.6905	0.05577	0.08077	17.932	12.381	**15**
16	1.485	0.6736	0.05160	0.07660	19.380	13.055	**16**
17	1.522	0.6572	0.04793	0.07293	20.865	13.712	**17**
18	1.560	0.6412	0.04467	0.06967	22.386	14.353	**18**
19	1.599	0.6255	0.04176	0.06676	23.946	14.979	**19**
20	1.639	0.6103	0.03915	0.06415	25.545	15.589	**20**
21	1.680	0.5954	0.03679	0.06179	27.183	16.185	**21**
22	1.722	0.5809	0.03465	0.05965	28.863	16.765	**22**
23	1.765	0.5667	0.03270	0.05770	30.584	17.332	**23**
24	1.809	0.5529	0.03091	0.05591	32.349	17.885	**24**
25	1.854	0.5394	0.02928	0.05428	34.158	18.424	**25**
26	1.900	0.5262	0.02777	0.05277	36.012	18.951	**26**
27	1.948	0.5134	0.02638	0.05138	37.912	19.464	**27**
28	1.996	0.5009	0.02509	0.05009	39.860	19.965	**28**
29	2.046	0.4887	0.02389	0.04889	41.856	20.454	**29**
30	2.098	0.4767	0.02278	0.04778	43.903	20.930	**30**
31	2.150	0.4651	0.02174	0.04674	46.000	21.395	**31**
32	2.204	0.4538	0.02077	0.04577	48.150	21.849	**32**
33	2.259	0.4427	0.01986	0.04486	50.354	22.292	**33**
34	2.315	0.4319	0.01901	0.04401	52.613	22.724	**34**
35	2.373	0.4214	0.01821	0.04321	54.928	23.145	**35**
40	2.685	0.3724	0.01484	0.03984	67.403	25.103	**40**
45	3.038	0.3292	0.01227	0.03727	81.516	26.833	**45**
50	3.437	0.2909	0.01026	0.03526	97.484	28.362	**50**
55	3.889	0.2572	0.00865	0.03365	115.551	29.714	**55**
60	4.400	0.2273	0.00735	0.03235	135.992	30.909	**60**
65	4.978	0.2009	0.00628	0.03128	159.118	31.965	**65**
70	5.632	0.1776	0.00540	0.03040	185.284	32.898	**70**
75	6.372	0.1569	0.00465	0.02965	214.888	33.723	**75**
80	7.210	0.1387	0.00403	0.02903	248.383	34.452	**80**
85	8.157	0.1226	0.00349	0.02849	286.279	35.096	**85**
90	9.229	0.1084	0.00304	0.02804	329.154	35.666	**90**
95	10.442	0.0958	0.00265	0.02765	377.664	36.169	**95**
100	11.814	0.0846	0.00231	0.02731	432.549	36.614	**100**

3% COMPOUND INTEREST TABLE

n	U	V	W	X	Y	Z	n
1	1.030	0.9709	1.00000	1.03000	1.000	0.971	1
2	1.061	0.9426	0.49261	0.52261	2.030	1.913	2
3	1.093	0.9151	0.32353	0.35353	3.091	2.829	3
4	1.126	0.8885	0.23903	0.26903	4.184	3.717	4
5	1.159	0.8626	0.18835	0.21835	5.309	4.580	5
6	1.194	0.8375	0.15460	0.18460	6.468	5.417	6
7	1.230	0.8131	0.13051	0.16051	7.662	6.230	7
8	1.267	0.7894	0.11246	0.14246	8.892	7.020	8
9	1.305	0.7664	0.09843	0.12843	10.159	7.786	9
10	1.344	0.7441	0.08723	0.11723	11.464	8.530	10
11	1.384	0.7224	0.07808	0.10808	12.808	9.253	11
12	1.426	0.7014	0.07046	0.10046	14.192	9.954	12
13	1.469	0.6810	0.06403	0.09403	15.618	10.635	13
14	1.513	0.6611	0.05853	0.08853	17.086	11.296	14
15	1.558	0.6419	0.05377	0.08377	18.599	11.938	15
16	1.605	0.6232	0.04961	0.07961	20.157	12.561	16
17	1.653	0.6050	0.04595	0.07595	21.762	13.166	17
18	1.702	0.5874	0.04271	0.07271	23.414	13.754	18
19	1.754	0.5703	0.03981	0.06981	25.117	14.324	19
20	1.806	0.5537	0.03722	0.06722	26.870	14.877	20
21	1.860	0.5375	0.03487	0.06487	28.676	15.415	21
22	1.916	0.5219	0.03275	0.06275	30.537	15.937	22
23	1.974	0.5067	0.03081	0.06081	32.453	16.444	23
24	2.033	0.4919	0.02905	0.05905	34.426	16.936	24
25	2.094	0.4776	0.02743	0.05743	36.459	17.413	25
26	2.157	0.4637	0.02594	0.05594	38.553	17.877	26
27	2.221	0.4502	0.02456	0.05456	40.710	18.327	27
28	2.288	0.4371	0.02329	0.05329	42.931	18.764	28
29	2.357	0.4243	0.02211	0.05211	45.219	19.188	29
30	2.427	0.4120	0.02102	0.05102	47.575	19.600	30
31	2.500	0.4000	0.02000	0.05000	50.003	20.000	31
32	2.575	0.3883	0.01905	0.04905	52.503	20.389	32
33	2.652	0.3770	0.01816	0.04816	55.078	20.766	33
34	2.732	0.3660	0.01732	0.04732	57.730	21.132	34
35	2.814	0.3554	0.01654	0.04654	60.462	21.487	35
40	3.262	0.3066	0.01326	0.04326	75.401	23.115	40
45	3.782	0.2644	0.01079	0.04079	92.720	24.519	45
50	4.384	0.2281	0.00887	0.03887	112.797	25.730	50
55	5.082	0.1968	0.00735	0.03735	136.072	26.774	55
60	5.892	0.1697	0.00613	0.03613	163.053	27.676	60
65	6.830	0.1464	0.00515	0.03515	194.333	28.453	65
70	7.918	0.1263	0.00434	0.03434	230.594	29.123	70
75	9.179	0.1089	0.00367	0.03367	272.631	29.702	75
80	10.641	0.0940	0.00311	0.03311	321.363	30.201	80
85	12.336	0.0811	0.00265	0.03265	377.857	30.631	85
90	14.300	0.0699	0.00226	0.03226	443.349	31.002	90
95	16.578	0.0603	0.00193	0.03193	519.272	31.323	95
100	19.219	0.0520	0.00165	0.03165	607.288	31.599	100

3.5% COMPOUND INTEREST TABLE

n	U	V	W	X	Y	Z	n
1	1.035	0.9662	1.00000	1.03500	1.000	0.966	1
2	1.071	0.9335	0.49140	0.52640	2.035	1.900	2
3	1.109	0.9019	0.32193	0.35693	3.106	2.802	3
4	1.148	0.8714	0.23725	0.27225	4.215	3.673	4
5	1.188	0.8420	0.18648	0.22148	5.362	4.515	5
6	1.229	0.8135	0.15267	0.18767	6.550	5.329	6
7	1.272	0.7860	0.12854	0.16354	7.779	6.115	7
8	1.317	0.7594	0.11048	0.14548	9.052	6.874	8
9	1.363	0.7337	0.09645	0.13145	10.368	7.608	9
10	1.411	0.7089	0.08524	0.12024	11.731	8.317	10
11	1.460	0.6849	0.07609	0.11109	13.142	9.002	11
12	1.511	0.6618	0.06848	0.10348	14.602	9.663	12
13	1.564	0.6394	0.06206	0.09706	16.113	10.303	13
14	1.619	0.6178	0.05657	0.09157	17.677	10.921	14
15	1.675	0.5969	0.05183	0.08683	19.296	11.517	15
16	1.734	0.5767	0.04768	0.08268	20.971	12.094	16
17	1.795	0.5572	0.04404	0.07904	22.705	12.651	17
18	1.857	0.5384	0.04082	0.07582	24.500	13.190	18
19	1.923	0.5202	0.03794	0.07294	26.357	13.710	19
20	1.990	0.5026	0.03536	0.07036	28.280	14.212	20
21	2.059	0.4856	0.03304	0.06804	30.269	14.698	21
22	2.132	0.4692	0.03093	0.06593	32.329	15.167	22
23	2.206	0.4533	0.02902	0.06402	34.460	15.620	23
24	2.283	0.4380	0.02727	0.06227	36.667	16.058	24
25	2.363	0.4231	0.02567	0.06067	38.950	16.482	25
26	2.446	0.4088	0.02421	0.05921	41.313	16.890	26
27	2.532	0.3950	0.02285	0.05785	43.759	17.285	27
28	2.620	0.3817	0.02160	0.05660	46.291	17.667	28
29	2.712	0.3687	0.02045	0.05545	48.911	18.036	29
30	2.807	0.3563	0.01937	0.05437	51.623	18.392	30
31	2.905	0.3442	0.01837	0.05337	54.429	18.736	31
32	3.007	0.3326	0.01744	0.05244	57.335	19.069	32
33	3.112	0.3213	0.01657	0.05157	60.341	19.390	33
34	3.221	0.3105	0.01576	0.05076	63.453	19.701	34
35	3.334	0.3000	0.01500	0.05000	66.674	20.001	35
40	3.959	0.2526	0.01183	0.04683	84.550	21.355	40
45	4.702	0.2127	0.00945	0.04445	105.782	22.495	45
50	5.585	0.1791	0.00763	0.04263	130.998	23.456	50
55	6.633	0.1508	0.00621	0.04121	160.947	24.264	55
60	7.878	0.1269	0.00509	0.04009	196.517	24.945	60
65	9.357	0.1069	0.00419	0.03919	238.763	25.518	65
70	11.113	0.0900	0.00346	0.03846	288.938	26.000	70
75	13.199	0.0758	0.00287	0.03787	348.530	26.407	75
80	15.676	0.0638	0.00238	0.03738	419.307	26.749	80
85	18.618	0.0537	0.00199	0.03699	503.367	27.037	85
90	22.112	0.0452	0.00166	0.03666	603.205	27.279	90
95	26.262	0.0381	0.00139	0.03639	721.781	27.484	95
100	31.191	0.0321	0.00116	0.03616	862.612	27.655	100

4% COMPOUND INTEREST TABLE

n	U	V	W	X	Y	Z	n
1	1.040	0.9615	1.00000	1.04000	1.000	0.962	1
2	1.082	0.9246	0.49020	0.53020	2.040	1.886	2
3	1.125	0.8890	0.32035	0.36035	3.122	2.775	3
4	1.170	0.8548	0.23549	0.27549	4.246	3.630	4
5	1.217	0.8219	0.18463	0.22463	5.416	4.452	5
6	1.265	0.7903	0.15076	0.19076	6.633	5.242	6
7	1.316	0.7599	0.12661	0.16661	7.898	6.002	7
8	1.369	0.7307	0.10853	0.14853	9.214	6.733	8
9	1.423	0.7026	0.09449	0.13449	10.583	7.435	9
10	1.480	0.6756	0.08329	0.12329	12.006	8.111	10
11	1.539	0.6496	0.07415	0.11415	13.486	8.760	11
12	1.601	0.6246	0.06655	0.10655	15.026	9.385	12
13	1.665	0.6006	0.06014	0.10014	16.627	9.986	13
14	1.732	0.5775	0.05467	0.09467	18.292	10.563	14
15	1.801	0.5553	0.04994	0.08994	20.024	11.118	15
16	1.873	0.5339	0.04582	0.08582	21.825	11.652	16
17	1.948	0.5134	0.04220	0.08220	23.698	12.166	17
18	2.026	0.4936	0.03899	0.07899	25.645	12.659	18
19	2.107	0.4746	0.03614	0.07614	27.671	13.134	19
20	2.191	0.4564	0.03358	0.07358	29.778	13.590	20
21	2.279	0.4388	0.03128	0.07128	31.969	14.029	21
22	2.370	0.4220	0.02920	0.06920	34.248	14.451	22
23	2.465	0.4057	0.02731	0.06731	36.618	14.857	23
24	2.563	0.3901	0.02559	0.06559	39.083	15.247	24
25	2.666	0.3751	0.02401	0.06401	41.646	15.622	25
26	2.772	0.3607	0.02257	0.06257	44.312	15.983	26
27	2.883	0.3468	0.02124	0.06124	47.084	16.330	27
28	2.999	0.3335	0.02001	0.06001	49.968	16.663	28
29	3.119	0.3207	0.01888	0.05888	52.966	16.984	29
30	3.243	0.3083	0.01783	0.05783	56.085	17.292	30
31	3.373	0.2965	0.01686	0.05686	59.328	17.588	31
32	3.508	0.2851	0.01595	0.05595	62.701	17.874	32
33	3.648	0.2741	0.01510	0.05510	66.210	18.148	33
34	3.794	0.2636	0.01431	0.05431	69.858	18.411	34
35	3.946	0.2534	0.01358	0.05358	73.652	18.665	35
40	4.801	0.2083	0.01052	0.05052	95.026	19.793	40
45	5.841	0.1712	0.00826	0.04826	121.029	20.720	45
50	7.107	0.1407	0.00655	0.04655	152.667	21.482	50
55	8.646	0.1157	0.00523	0.04523	191.159	22.109	55
60	10.520	0.0951	0.00420	0.04420	237.991	22.623	60
65	12.799	0.0781	0.00339	0.04339	294.968	23.047	65
70	15.572	0.0642	0.00275	0.04275	364.290	23.395	70
75	18.945	0.0528	0.00223	0.04223	448.631	23.680	75
80	23.050	0.0434	0.00181	0.04181	551.245	23.915	80
85	28.044	0.0357	0.00148	0.04148	676.090	24.109	85
90	34.119	0.0293	0.00121	0.04121	827.983	24.267	90
95	41.511	0.0241	0.00099	0.04099	1012.785	24.398	95
100	50.505	0.0198	0.00081	0.04081	1237.624	24.505	100

4.5% COMPOUND INTEREST TABLE

n	U	V	W	X	Y	Z	n
1	1.045	0.9569	1.00000	1.04500	1.000	0.957	1
2	1.092	0.9157	0.48900	0.53400	2.045	1.873	2
3	1.141	0.8763	0.31877	0.36377	3.137	2.749	3
4	1.193	0.8386	0.23374	0.27874	4.278	3.588	4
5	1.246	0.8025	0.18279	0.22779	5.471	4.390	5
6	1.302	0.7679	0.14888	0.19388	6.717	5.158	6
7	1.361	0.7348	0.12470	0.16970	8.019	5.893	7
8	1.422	0.7032	0.10661	0.15161	9.380	6.596	8
9	1.486	0.6729	0.09257	0.13757	10.802	7.269	9
10	1.553	0.6439	0.08138	0.12638	12.288	7.913	10
11	1.623	0.6162	0.07225	0.11725	13.841	8.529	11
12	1.696	0.5897	0.06467	0.10967	15.464	9.119	12
13	1.772	0.5643	0.05828	0.10328	17.160	9.683	13
14	1.852	0.5400	0.05282	0.09782	18.932	10.223	14
15	1.935	0.5167	0.04811	0.09311	20.784	10.740	15
16	2.022	0.4945	0.04402	0.08902	22.719	11.234	16
17	2.113	0.4732	0.04042	0.08542	24.742	11.707	17
18	2.208	0.4528	0.03724	0.08224	26.855	12.160	18
19	2.308	0.4333	0.03441	0.07941	29.064	12.593	19
20	2.412	0.4146	0.03188	0.07688	31.371	13.008	20
21	2.520	0.3968	0.02960	0.07460	33.783	13.405	21
22	2.634	0.3797	0.02755	0.07255	36.303	13.784	22
23	2.752	0.3634	0.02568	0.07068	38.937	14.148	23
24	2.876	0.3477	0.02399	0.06899	41.689	14.495	24
25	3.005	0.3327	0.02244	0.06744	44.565	14.828	25
26	3.141	0.3184	0.02102	0.06602	47.571	15.147	26
27	3.282	0.3047	0.01972	0.06472	50.711	15.451	27
28	3.430	0.2916	0.01852	0.06352	53.993	15.743	28
29	3.584	0.2790	0.01741	0.06241	57.423	16.022	29
30	3.745	0.2670	0.01639	0.06139	61.007	16.289	30
31	3.914	0.2555	0.01544	0.06044	64.752	16.544	31
32	4.090	0.2445	0.01456	0.05956	68.666	16.789	32
33	4.274	0.2340	0.01374	0.05874	72.756	17.023	33
34	4.466	0.2239	0.01298	0.05798	77.030	17.247	34
35	4.667	0.2143	0.01227	0.05727	81.497	17.461	35
40	5.816	0.1719	0.00934	0.05434	107.030	18.402	40
45	7.248	0.1380	0.00720	0.05220	138.850	19.156	45
50	9.033	0.1107	0.00560	0.05060	178.503	19.762	50
55	11.256	0.0888	0.00439	0.04939	227.918	20.248	55
60	14.027	0.0713	0.00345	0.04845	289.498	20.638	60
65	17.481	0.0572	0.00273	0.04773	366.238	20.951	65
70	21.784	0.0459	0.00217	0.04717	461.870	21.202	70
75	27.147	0.0368	0.00172	0.04672	581.044	21.404	75
80	33.830	0.0296	0.00137	0.04637	729.558	21.565	80
85	42.158	0.0237	0.00109	0.04609	914.632	21.695	85
90	52.537	0.0190	0.00087	0.04587	1145.269	21.799	90
95	65.471	0.0153	0.00070	0.04570	1432.684	21.883	95
100	81.589	0.0123	0.00056	0.04556	1790.856	21.950	100

5% COMPOUND INTEREST TABLE

n	U	V	W	X	Y	Z	n
1	1.050	0.9524	1.00000	1.05000	1.000	0.952	1
2	1.103	0.9070	0.48780	0.53780	2.050	1.859	2
3	1.158	0.8638	0.31721	0.36721	3.153	2.723	3
4	1.216	0.8227	0.23201	0.28201	4.310	3.546	4
5	1.276	0.7835	0.18097	0.23097	5.526	4.329	5
6	1.340	0.7462	0.14702	0.19702	6.802	5.076	6
7	1.407	0.7107	0.12282	0.17282	8.142	5.786	7
8	1.477	0.6768	0.10472	0.15472	9.549	6.463	8
9	1.551	0.6446	0.09069	0.14069	11.027	7.108	9
10	1.629	0.6139	0.07950	0.12950	12.578	7.722	10
11	1.710	0.5847	0.07039	0.12039	14.207	8.306	11
12	1.796	0.5568	0.06283	0.11283	15.917	8.863	12
13	1.886	0.5303	0.05646	0.10646	17.713	9.394	13
14	1.980	0.5051	0.05102	0.10102	19.599	9.899	14
15	2.079	0.4810	0.04634	0.09634	21.579	10.380	15
16	2.183	0.4581	0.04227	0.09227	23.657	10.838	16
17	2.292	0.4363	0.03870	0.08870	25.840	11.274	17
18	2.407	0.4155	0.03555	0.08555	28.132	11.690	18
19	2.527	0.3957	0.03275	0.08275	30.539	12.085	19
20	2.653	0.3769	0.03024	0.08024	33.066	12.462	20
21	2.786	0.3589	0.02800	0.07800	35.719	12.821	21
22	2.925	0.3418	0.02597	0.07597	38.505	13.163	22
23	3.072	0.3256	0.02414	0.07414	41.430	13.489	23
24	3.225	0.3101	0.02247	0.07247	44.502	13.799	24
25	3.386	0.2953	0.02095	0.07095	47.727	14.094	25
26	3.556	0.2812	0.01956	0.06956	51.113	14.375	26
27	3.733	0.2678	0.01829	0.06829	54.669	14.643	27
28	3.920	0.2551	0.01712	0.06712	58.403	14.898	28
29	4.116	0.2429	0.01605	0.06605	62.323	15.141	29
30	4.322	0.2314	0.01505	0.06505	66.439	15.372	30
31	4.538	0.2204	0.01413	0.06413	70.761	15.593	31
32	4.765	0.2099	0.01328	0.06328	75.299	15.803	32
33	5.003	0.1999	0.01249	0.06249	80.064	16.003	33
34	5.253	0.1904	0.01176	0.06176	85.067	16.193	34
35	5.516	0.1813	0.01107	0.06107	90.320	16.374	35
40	7.040	0.1420	0.00828	0.05828	120.800	17.159	40
45	8.985	0.1113	0.00626	0.05626	159.700	17.774	45
50	11.467	0.0872	0.00478	0.05478	209.348	18.256	50
55	14.636	0.0683	0.00367	0.05367	272.713	18.633	55
60	18.679	0.0535	0.00283	0.05283	353.584	18.929	60
65	23.840	0.0419	0.00219	0.05219	456.798	19.161	65
70	30.426	0.0329	0.00170	0.05170	588.529	19.343	70
75	38.833	0.0258	0.00132	0.05132	756.654	19.485	75
80	49.561	0.0202	0.00103	0.05103	971.229	19.596	80
85	63.254	0.0158	0.00080	0.05080	1245.087	19.684	85
90	80.730	0.0124	0.00063	0.05063	1594.607	19.752	90
95	103.035	0.0097	0.00049	0.05049	2040.694	19.806	95
100	131.501	0.0076	0.00038	0.05038	2610.025	19.848	100

5.5% COMPOUND INTEREST TABLE

n	U	V	W	X	Y	Z	n
1	1.055	0.9479	1.00000	1.05500	1.000	0.948	1
2	1.113	0.8985	0.48662	0.54162	2.055	1.846	2
3	1.174	0.8516	0.31565	0.37065	3.168	2.698	3
4	1.239	0.8072	0.23029	0.28529	4.342	3.505	4
5	1.307	0.7651	0.17918	0.23418	5.581	4.270	5
6	1.379	0.7252	0.14518	0.20018	6.888	4.996	6
7	1.455	0.6874	0.12096	0.17596	8.267	5.683	7
8	1.535	0.6516	0.10286	0.15786	9.722	6.335	8
9	1.619	0.6176	0.08884	0.14384	11.256	6.952	9
10	1.708	0.5854	0.07767	0.13267	12.875	7.538	10
11	1.802	0.5549	0.06857	0.12357	14.583	8.093	11
12	1.901	0.5260	0.06103	0.11603	16.386	8.619	12
13	2.006	0.4986	0.05468	0.10968	18.287	9.117	13
14	2.116	0.4726	0.04928	0.10428	20.293	9.590	14
15	2.232	0.4479	0.04463	0.09963	22.409	10.038	15
16	2.355	0.4246	0.04058	0.09558	24.641	10.462	16
17	2.485	0.4024	0.03704	0.09204	26.996	10.865	17
18	2.621	0.3815	0.03392	0.08892	29.481	11.246	18
19	2.766	0.3616	0.03115	0.08615	32.103	11.608	19
20	2.918	0.3427	0.02868	0.08368	34.868	11.950	20
21	3.078	0.3249	0.02646	0.08146	37.786	12.275	21
22	3.248	0.3079	0.02447	0.07947	40.864	12.583	22
23	3.426	0.2919	0.02267	0.07767	44.112	12.875	23
24	3.615	0.2767	0.02104	0.07604	47.538	13.152	24
25	3.813	0.2622	0.01955	0.07455	51.153	13.414	25
26	4.023	0.2486	0.01819	0.07319	54.966	13.662	26
27	4.244	0.2356	0.01695	0.07195	58.989	13.898	27
28	4.478	0.2233	0.01581	0.07081	63.234	14.121	28
29	4.724	0.2117	0.01477	0.06977	67.711	14.333	29
30	4.984	0.2006	0.01381	0.06881	72.435	14.534	30
31	5.258	0.1902	0.01292	0.06792	77.419	14.724	31
32	5.547	0.1803	0.01210	0.06710	82.677	14.904	32
33	5.852	0.1709	0.01133	0.06633	88.225	15.075	33
34	6.174	0.1620	0.01063	0.06563	94.077	15.237	34
35	6.514	0.1535	0.00997	0.06497	100.251	15.391	35
40	8.513	0.1175	0.00732	0.06232	136.606	16.046	40
45	11.127	0.0899	0.00543	0.06043	184.119	16.548	45
50	14.542	0.0688	0.00406	0.05906	246.217	16.932	50
55	19.006	0.0526	0.00305	0.05805	327.377	17.225	55
60	24.840	0.0403	0.00231	0.05731	433.450	17.450	60
65	32.465	0.0308	0.00175	0.05675	572.083	17.622	65
70	42.430	0.0236	0.00133	0.05633	753.271	17.753	70
75	55.454	0.0180	0.00101	0.05601	990.076	17.854	75
80	72.476	0.0138	0.00077	0.05577	1299.571	17.931	80
85	94.724	0.0106	0.00059	0.05559	1704.069	17.990	85
90	123.800	0.0081	0.00045	0.05545	2232.731	18.035	90
95	161.802	0.0062	0.00034	0.05534	2923.671	18.069	95
100	211.469	0.0047	0.00026	0.05526	3826.702	18.096	100

6% COMPOUND INTEREST TABLE

n	U	V	W	X	Y	Z	n
1	1.060	0.9434	1.00000	1.06000	1.000	0.943	1
2	1.124	0.8900	0.48544	0.54544	2.060	1.833	2
3	1.191	0.8396	0.31411	0.37411	3.184	2.673	3
4	1.262	0.7921	0.22859	0.28859	4.375	3.465	4
5	1.338	0.7473	0.17740	0.23740	5.637	4.212	5
6	1.419	0.7050	0.14336	0.20336	6.975	4.917	6
7	1.504	0.6651	0.11914	0.17914	8.394	5.582	7
8	1.594	0.6274	0.10104	0.16104	9.897	6.210	8
9	1.689	0.5919	0.08702	0.14702	11.491	6.802	9
10	1.791	0.5584	0.07587	0.13587	13.181	7.360	10
11	1.898	0.5268	0.06679	0.12679	14.972	7.887	11
12	2.012	0.4970	0.05928	0.11928	16.870	8.384	12
13	2.133	0.4688	0.05296	0.11296	18:882	8.853	13
14	2.261	0.4423	0.04758	0.10758	21.015	9.295	14
15	2.397	0.4173	0.04296	0.10296	23.276	9.712	15
16	2.540	0.3936	0.03895	0.09895	25.673	10.106	16
17	2.693	0.3714	0.03544	0.09544	28.213	10.477	17
18	2.854	0.3503	0.03236	0.09236	30.906	10.828	18
19	3.026	0.3305	0.02962	0.08962	33.760	11.158	19
20	3.207	0.3118	0.02718	0.08718	36.786	11.470	20
21	3.400	0.2942	0.02500	0.08500	39.993	11.764	21
22	3.604	0.2775	0.02305	0.08305	43.392	12.042	22
23	3.820	0.2618	0.02128	0.08128	46.996	12.303	23
24	4.049	0.2470	0.01968	0.07968	50.816	12.550	24
25	4.292	0.2330	0.01823	0.07823	54.865	12.783	25
26	4.549	0.2198	0.01690	0.07690	59.156	13.003	26
27	4.822	0.2074	0.01570	0.07570	63.706	13.211	27
28	5.112	0.1956	0.01459	0.07459	68.528	13.406	28
29	5.418	0.1846	0.01358	0.07358	73.640	13.591	29
30	5.743	0.1741	0.01265	0.07265	79.058	13.765	30
31	6.088	0.1643	0.01179	0.07179	84.802	13.929	31
32	6.453	0.1550	0.01100	0.07100	90.890	14.084	32
33	6.841	0.1462	0.01027	0.07027	97.343	14.230	33
34	7.251	0.1379	0.00960	0.06960	104.184	14.368	34
35	7.686	0.1301	0.00897	0.06897	111.435	14.498	35
40	10.286	0.0972	0.00646	0.06646	154.762	15.046	40
45	13.765	0.0727	0.00470	0.06470	212.744	15.456	45
50	18.420	0.0543	0.00344	0.06344	290.336	15.762	50
55	24.650	0.0406	0.00254	0.06254	394.172	15.991	55
60	32.988	0.0303	0.00188	0.06188	533.128	16.161	60
65	44.145	0.0227	0.00139	0.06139	719.083	16.289	65
70	59.076	0.0169	0.00103	0.06103	967.932	16.385	70
75	79.057	0.0126	0.00077	0.06077	1300.949	16.456	75
80	105.796	0.0095	0.00057	0.06057	1746.600	16.509	80
85	141.579	0.0071	0.00043	0.06043	2342.982	16.549	85
90	189.465	0.0053	0.00032	0.06032	3141.075	16.579	90
95	253.546	0.0039	0.00024	0.06024	4209.104	16.601	95
100	339.302	0.0029	0.00018	0.06018	5638.368	16.618	100

7% COMPOUND INTEREST TABLE

n	U	V	W	X	Y	Z	n
1	1.070	0.9346	1.00000	1.07000	1.000	0.935	**1**
2	1.145	0.8734	0.48309	0.55309	2.070	1.808	**2**
3	1.225	0.8163	0.31105	0.38105	3.215	2.624	**3**
4	1.311	0.7629	0.22523	0.29523	4.440	3.387	**4**
5	1.403	0.7130	0.17389	0.24389	5.751	4.100	**5**
6	1.501	0.6663	0.13980	0.20980	7.153	4.767	**6**
7	1.606	0.6227	0.11555	0.18555	8.654	5.389	**7**
8	1.718	0.5820	0.09747	0.16747	10.260	5.971	**8**
9	1.838	0.5439	0.08349	0.15349	11.978	6.515	**9**
10	1.967	0.5083	0.07238	0.14238	13.816	7.024	**10**
11	2.105	0.4751	0.06336	0.13336	15.784	7.499	**11**
12	2.252	0.4440	0.05590	0.12590	17.888	7.943	**12**
13	2.410	0.4150	0.04965	0.11965	20.141	8.358	**13**
14	2.579	0.3878	0.04434	0.11434	22.550	8.745	**14**
15	2.759	0.3624	0.03979	0.10979	25.129	9.108	**15**
16	2.952	0.3387	0.03586	0.10586	27.888	9.447	**16**
17	3.159	0.3166	0.03243	0.10243	30.840	9.763	**17**
18	3.380	0.2959	0.02941	0.09941	33.999	10.059	**18**
19	3.617	0.2765	0.02675	0.09675	37.379	10.336	**19**
20	3.870	0.2584	0.02439	0.09439	40.995	10.594	**20**
21	4.141	0.2415	0.02229	0.09229	44.865	10.836	**21**
22	4.430	0.2257	0.02041	0.09041	49.006	11.061	**22**
23	4.741	0.2109	0.01871	0.08871	53.436	11.272	**23**
24	5.072	0.1971	0.01719	0.08719	58.177	11.469	**24**
25	5.427	0.1842	0.01581	0.08581	63.249	11.654	**25**
26	5.807	0.1722	0.01456	0.08456	68.676	11.826	**26**
27	6.214	0.1609	0.01343	0.08343	74.484	11.987	**27**
28	6.649	0.1504	0.01239	0.08239	80.698	12.137	**28**
29	7.114	0.1406	0.01145	0.08145	87.347	12.278	**29**
30	7.612	0.1314	0.01059	0.08059	94.461	12.409	**30**
31	8.145	0.1228	0.00980	0.07980	102.073	12.532	**31**
32	8.715	0.1147	0.00907	0.07907	110.218	12.647	**32**
33	9.325	0.1072	0.00841	0.07841	118.933	12.754	**33**
34	9.978	0.1002	0.00780	0.07780	128.259	12.854	**34**
35	10.677	0.0937	0.00723	0.07723	138.237	12.948	**35**
40	14.974	0.0668	0.00501	0.07501	199.635	13.332	**40**
45	21.002	0.0476	0.00350	0.07350	285.749	13.606	**45**
50	29.457	0.0339	0.00246	0.07246	406.529	13.801	**50**
55	41.315	0.0242	0.00174	0.07174	575.929	13.940	**55**
60	57.946	0.0173	0.00123	0.07123	813.520	14.039	**60**
65	81.273	0.0123	0.00087	0.07087	1146.755	14.110	**65**
70	113.989	0.0088	0.00062	0.07062	1614.134	14.160	**70**
75	159.876	0.0063	0.00044	0.07044	2269.657	14.196	**75**
80	224.234	0.0045	0.00031	0.07031	3189.063	14.222	**80**
85	314.500	0.0032	0.00022	0.07022	4478.576	14.240	**85**
90	441.103	0.0023	0.00016	0.07016	6287.185	14.253	**90**
95	618.670	0.0016	0.00011	0.07011	8823.854	14.263	**95**
100	867.716	0.0012	0.00008	0.07008	12381.662	14.269	**100**

8% COMPOUND INTEREST TABLE

n	U	V	W	X	Y	Z	n
1	1.080	0.9259	1.00000	1.08000	1.000	0.926	1
2	1.166	0.8573	0.48077	0.56077	2.080	1.783	2
3	1.260	0.7938	0.30803	0.38803	3.246	2.577	3
4	1.360	0.7350	0.22192	0.30192	4.506	3.312	4
5	1.469	0.6806	0.17046	0.25046	5.867	3.993	5
6	1.587	0.6302	0.13632	0.21632	7.336	4.623	6
7	1.714	0.5835	0.11207	0.19207	8.923	5.206	7
8	1.851	0.5403	0.09401	0.17401	10.637	5.747	8
9	1.999	0.5002	0.08008	0.16008	12.488	6.247	9
10	2.159	0.4632	0.06903	0.14903	14.487	6.710	10
11	2.332	0.4289	0.06008	0.14008	16.645	7.139	11
12	2.518	0.3971	0.05270	0.13270	18.977	7.536	12
13	2.720	0.3677	0.04652	0.12652	21.495	7.904	13
14	2.937	0.3405	0.04130	0.12130	24.215	8.244	14
15	3.172	0.3152	0.03683	0.11683	27.152	8.559	15
16	3.426	0.2919	0.03298	0.11298	30.324	8.851	16
17	3.700	0.2703	0.02963	0.10963	33.750	9.122	17
18	3.996	0.2502	0.02670	0.10670	37.450	9.372	18
19	4.316	0.2317	0.02413	0.10413	41.446	9.604	19
20	4.661	0.2145	0.02185	0.10185	45.762	9.818	20
21	5.034	0.1987	0.01983	0.09983	50.423	10.017	21
22	5.437	0.1839	0.01803	0.09803	55.457	10.201	22
23	5.871	0.1703	0.01642	0.09642	60.893	10.371	23
24	6.341	0.1577	0.01498	0.09498	66.765	10.529	24
25	6.848	0.1460	0.01368	0.09368	73.106	10.675	25
26	7.396	0.1352	0.01251	0.09251	79.954	10.810	26
27	7.988	0.1252	0.01145	0.09145	87.351	10.935	27
28	8.627	0.1159	0.01049	0.09049	95.339	11.051	28
29	9.317	0.1073	0.00962	0.08962	103.966	11.158	29
30	10.063	0.0994	0.00883	0.08883	113.283	11.258	30
31	10.868	0.0920	0.00811	0.08811	123.346	11.350	31
32	11.737	0.0852	0.00745	0.08745	134.214	11.435	32
33	12.676	0.0789	0.00685	0.08685	145.951	11.514	33
34	13.690	0.0730	0.00630	0.08630	158.627	11.587	34
35	14.785	0.0676	0.00580	0.08580	172.317	11.655	35
40	21.725	0.0460	0.00386	0.08386	259.057	11.925	40
45	31.920	0.0313	0.00259	0.08259	386.506	12.108	45
50	46.902	0.0213	0.00174	0.08174	573.770	12.233	50
55	68.914	0.0145	0.00118	0.08118	848.923	12.319	55
60	101.257	0.0099	0.00080	0.08080	1253.213	12.377	60
65	148.780	0.0067	0.00054	0.08054	1847.248	12.416	65
70	218.606	0.0046	0.00037	0.08037	2720.080	12.443	70
75	321.205	0.0031	0.00025	0.08025	4002.557	12.461	75
80	471.955	0.0021	0.00017	0.08017	5886.935	12.474	80
85	693.456	0.0014	0.00012	0.08012	8655.706	12.482	85
90	1018.915	0.0010	0.00008	0.08008	12723.939	12.488	90
95	1497.121	0.0007	0.00005	0.08005	18701.507	12.492	95
100	2199.761	0.0005	0.00004	0.08004	27484.516	12.494	100

10% COMPOUND INTEREST TABLE

n	U	V	W	X	Y	Z	n
1	1.100	0.9091	1.00000	1.10000	1.000	0.909	**1**
2	1.210	0.8264	0.47619	0.57619	2.100	1.736	**2**
3	1.331	0.7513	0.30211	0.40211	3.310	2.487	**3**
4	1.464	0.6830	0.21547	0.31547	4.641	3.170	**4**
5	1.611	0.6209	0.16380	0.26380	6.105	3.791	**5**
6	1.772	0.5645	0.12961	0.22961	7.716	4.355	**6**
7	1.949	0.5132	0.10541	0.20541	9.487	4.868	**7**
8	2.144	0.4665	0.08744	0.18744	11.436	5.335	**8**
9	2.358	0.4241	0.07364	0.17364	13.579	5.759	**9**
10	2.594	0.3855	0.06275	0.16275	15.937	6.144	**10**
11	2.853	0.3505	0.05396	0.15396	18.531	6.495	**11**
12	3.138	0.3186	0.04676	0.14676	21.384	6.814	**12**
13	3.452	0.2897	0.04078	0.14078	24.523	7.103	**13**
14	3.797	0.2633	0.03575	0.13575	27.975	7.367	**14**
15	4.177	0.2394	0.03147	0.13147	31.772	7.606	**15**
16	4.595	0.2176	0.02782	0.12782	35.950	7.824	**16**
17	5.054	0.1978	0.02466	0.12466	40.545	8.022	**17**
18	5.560	0.1799	0.02193	0.12193	45.599	8.201	**18**
19	6.116	0.1635	0.01955	0.11955	51.159	8.365	**19**
20	6.727	0.1486	0.01746	0.11746	57.275	8.514	**20**
21	7.400	0.1351	0.01562	0.11562	64.002	8.649	**21**
22	8.140	0.1228	0.01401	0.11401	71.403	8.772	**22**
23	8.954	0.1117	0.01257	0.11257	79.543	8.883	**23**
24	9.850	0.1015	0.01130	0.11130	88.497	8.985	**24**
25	10.835	0.0923	0.01017	0.11017	98.347	9.077	**25**
26	11.918	0.0839	0.00916	0.10916	109.182	9.161	**26**
27	13.110	0.0763	0.00826	0.10826	121.100	9.237	**27**
28	14.421	0.0693	0.00745	0.10745	134.210	9.307	**28**
29	15.863	0.0630	0.00673	0.10673	148.631	9.370	**29**
30	17.449	0.0573	0.00608	0.10608	164.494	9.427	**30**
31	19.194	0.0521	0.00550	0.10550	181.943	9.479	**31**
32	21.114	0.0474	0.00497	0.10497	201.138	9.526	**32**
33	23.225	0.0431	0.00450	0.10450	222.252	9.569	**33**
34	25.548	0.0391	0.00407	0.10407	245.477	9.609	**34**
35	28.102	0.0356	0.00369	0.10369	271.024	9.644	**35**
40	45.259	0.0221	0.00226	0.10226	442.593	9.779	**40**
45	72.890	0.0137	0.00139	0.10139	718.905	9.863	**45**
50	117.391	0.0085	0.00086	0.10086	1163.909	9.915	**50**
55	189.059	0.0053	0.00053	0.10053	1880.591	9.947	**55**
60	304.482	0.0033	0.00033	0.10033	3034.816	9.967	**60**
65	490.371	0.0020	0.00020	0.10020	4893.707	9.980	**65**
70	789.747	0.0013	0.00013	0.10013	7887.470	9.987	**70**
75	1271.895	0.0008	0.00008	0.10008	12708.954	9.992	**75**
80	2048.400	0.0005	0.00005	0.10005	20474.002	9.995	**80**
85	3298.969	0.0003	0.00003	0.10003	32979.690	9.997	**85**
90	5313.023	0.0002	0.00002	0.10002	53120.226	9.998	**90**
95	8556.676	0.0001	0.00001	0.10001	85556.760	9.999	**95**
100	13780.612	0.0001	0.00001	0.10001	137796.123	9.999	**100**

12% COMPOUND INTEREST TABLE

n	U	V	W	X	Y	Z	n
1	1.120	0.8929	1.00000	1.12000	1.000	0.893	**1**
2	1.254	0.7972	0.47170	0.59170	2.120	1.690	**2**
3	1.405	0.7118	0.29635	0.41635	3.374	2.402	**3**
4	1.574	0.6355	0.20923	0.32923	4.779	3.037	**4**
5	1.762	0.5674	0.15741	0.27741	6.353	3.605	**5**
6	1.974	0.5066	0.12323	0.24323	8.115	4.111	**6**
7	2.211	0.4523	0.09912	0.21912	10.089	4.564	**7**
8	2.476	0.4039	0.08130	0.20130	12.300	4.968	**8**
9	2.773	0.3606	0.06768	0.18768	14.776	5.328	**9**
10	3.106	0.3220	0.05698	0.17698	17.549	5.650	**10**
11	3.479	0.2875	0.04842	0.16842	20.655	5.938	**11**
12	3.896	0.2567	0.04144	0.16144	24.133	6.194	**12**
13	4.363	0.2292	0.03568	0.15568	28.029	6.424	**13**
14	4.887	0.2046	0.03087	0.15087	32.393	6.628	**14**
15	5.474	0.1827	0.02682	0.14682	37.280	6.811	**15**
16	6.130	0.1631	0.02339	0.14339	42.753	6.974	**16**
17	6.866	0.1456	0.02046	0.14046	48.884	7.120	**17**
18	7.690	0.1300	0.01794	0.13794	55.750	7.250	**18**
19	8.613	0.1161	0.01576	0.13576	63.440	7.366	**19**
20	9.646	0.1037	0.01388	0.13388	72.052	7.469	**20**
21	10.804	0.0926	0.01224	0.13224	81.699	7.562	**21**
22	12.100	0.0826	0.01081	0.13081	92.502	7.645	**22**
23	13.552	0.0738	0.00956	0.12956	104.603	7.718	**23**
24	15.179	0.0659	0.00846	0.12846	118.155	7.784	**24**
25	17.000	0.0588	0.00750	0.12750	133.334	7.843	**25**
26	19.040	0.0525	0.00665	0.12665	150.334	7.896	**26**
27	21.325	0.0469	0.00590	0.12590	169.374	7.943	**27**
28	23.884	0.0419	0.00524	0.12524	190.699	7.984	**28**
29	26.750	0.0374	0.00466	0.12466	214.582	8.022	**29**
30	29.960	0.0334	0.00414	0.12414	241.332	8.055	**30**
31	33.555	0.0298	0.00369	0.12369	271.292	8.085	**31**
32	37.582	0.0266	0.00328	0.12328	304.847	8.112	**32**
33	42.091	0.0238	0.00292	0.12292	342.429	8.135	**33**
34	47.142	0.0212	0.00260	0.12260	384.520	8.157	**34**
35	52.799	0.0189	0.00232	0.12232	431.663	8.176	**35**
40	93.051	0.0107	0.00130	0.12130	767.088	8.244	**40**
45	163.987	0.0061	0.00074	0.12074	1358.224	8.283	**45**
50	289.001	0.0035	0.00042	0.12042	2400.008	8.305	**50**
∞				0.12000		8.333	**∞**

15% COMPOUND INTEREST TABLE

n	U	V	W	X	Y	Z	n
1	1.150	0.8696	1.00000	1.15000	1.000	0.870	**1**
2	1.322	0.7561	0.46512	0.61512	2.150	1.626	**2**
3	1.521	0.6575	0.28798	0.43798	3.472	2.283	**3**
4	1.749	0.5718	0.20026	0.35027	4.993	2.855	**4**
5	2.011	0.4972	0.14832	0.29832	6.742	3.352	**5**
6	2.313	0.4323	0.11424	0.26424	8.754	3.784	**6**
7	2.660	0.3759	0.09036	0.24036	11.067	4.160	**7**
8	3.059	0.3269	0.07285	0.22285	13.727	4.487	**8**
9	3.518	0.2843	0.05957	0.20957	16.786	4.772	**9**
10	4.046	0.2472	0.04925	0.19925	20.304	5.019	**10**
11	4.652	0.2149	0.04107	0.19107	24.349	5.234	**11**
12	5.350	0.1869	0.03448	0.18448	29.002	5.421	**12**
13	6.153	0.1625	0.02911	0.17911	34.352	5.583	**13**
14	7.076	0.1413	0.02469	0.17469	40.505	5.724	**14**
15	8.137	0.1229	0.02102	0.17102	47.580	5.847	**15**
16	9.358	0.1069	0.01795	0.16795	55.717	5.954	**16**
17	10.761	0.0929	0.01537	0.16537	65.075	6.047	**17**
18	12.375	0.0808	0.01319	0.16319	75.836	6.128	**18**
19	14.232	0.0703	0.01134	0.16134	88.212	6.198	**19**
20	16.367	0.0611	0.00976	0.15976	102.443	6.259	**20**
21	18.821	0.0531	0.00842	0.15842	118.810	6.312	**21**
22	21.645	0.0462	0.00727	0.15727	137.631	6.359	**22**
23	24.891	0.0402	0.00628	0.15628	159.276	6.399	**23**
24	28.625	0.0349	0.00543	0.15543	184.167	6.434	**24**
25	32.919	0.0304	0.00470	0.15470	212.793	6.464	**25**
26	37.857	0.0264	0.00407	0.15407	245.711	6.491	**26**
27	43.535	0.0230	0.00353	0.15353	283.568	6.514	**27**
28	50.065	0.0200	0.00306	0.15306	327.103	6.534	**28**
29	57.575	0.0174	0.00265	0.15265	377.169	6.551	**29**
30	66.212	0.0151	0.00230	0.15230	434.744	6.566	**30**
31	76.143	0.0131	0.00200	0.15200	500.956	6.579	**31**
32	87.565	0.0114	0.00173	0.15173	577.099	6.591	**32**
33	100.700	0.0099	0.00150	0.15150	664.664	6.600	**33**
34	115.805	0.0086	0.00131	0.15131	765.364	6.609	**34**
35	133.175	0.0075	0.00113	0.15113	881.168	6.617	**35**
40	267.862	0.0037	0.00056	0.15056	1779.1	6.642	**40**
45	538.767	0.0019	0.00028	0.15028	3585.1	6.654	**45**
50	1083.652	0.0009	0.00014	0.15014	7217.7	6.661	**50**
∞				0.15000		6.667	**∞**

20% COMPOUND INTEREST TABLE

n	U	V	W	X	Y	Z	n
1	1.200	0.8333	1.00000	1.20000	1.000	0.833	**1**
2	1.440	0.6944	0.45455	0.65455	2.200	1.528	**2**
3	1.728	0.5787	0.27473	0.47473	3.640	2.106	**3**
4	2.074	0.4823	0.18629	0.38629	5.368	2.589	**4**
5	2.488	0.4019	0.13438	0.33438	7.442	2.991	**5**
6	2.986	0.3349	0.10071	0.30071	9.930	3.326	**6**
7	3.583	0.2791	0.07742	0.27742	12.916	3.605	**7**
8	4.300	0.2326	0.06061	0.26061	16.499	3.837	**8**
9	5.160	0.1938	0.04808	0.24808	20.799	4.031	**9**
10	6.192	0.1615	0.03852	0.23852	25.959	4.192	**10**
11	7.430	0.1346	0.03110	0.23110	32.150	4.327	**11**
12	8.916	0.1122	0.02526	0.22526	39.580	4.439	**12**
13	10.699	0.0935	0.02062	0.22062	48.497	4.533	**13**
14	12.839	0.0779	0.01689	0.21689	59.196	4.611	**14**
15	15.407	0.0649	0.01388	0.21388	72.035	4.675	**15**
16	18.488	0.0541	0.01144	0.21144	87.442	4.730	**16**
17	22.186	0.0451	0.00944	0.20944	105.931	4.775	**17**
18	26.623	0.0376	0.00781	0.20781	128.117	4.812	**18**
19	31.948	0.0313	0.00646	0.20646	154.740	4.844	**19**
20	38.338	0.0261	0.00536	0.20536	186.688	4.870	**20**
21	46.005	0.0217	0.00444	0.20444	225.025	4.891	**21**
22	55.206	0.0181	0.00369	0.20369	271.031	4.909	**22**
23	66.247	0.0151	0.00307	0.20307	326.237	4.925	**23**
24	79.497	0.0126	0.00255	0.20255	392.484	4.937	**24**
25	95.396	0.0105	0.00212	0.20212	471.981	4.948	**25**
26	114.475	0.0087	0.00176	0.20176	567.377	4.956	**26**
27	137.370	0.0073	0.00147	0.20147	681.852	4.964	**27**
28	164.845	0.0061	0.00122	0.20122	819.223	4.970	**28**
29	197.813	0.0051	0.00102	0.20102	984.067	4.975	**29**
30	237.376	0.0042	0.00085	0.20085	1181.881	4.979	**30**
31	284.851	0.0035	0.00070	0.20070	1419.257	4.982	**31**
32	341.822	0.0029	0.00059	0.20059	1704.108	4.985	**32**
33	410.186	0.0024	0.00049	0.20049	2045.930	4.988	**33**
34	492.223	0.0020	0.00041	0.20041	2456.116	4.990	**34**
35	590.668	0.0017	0.00034	0.20034	2948.339	4.992	**35**
40	1469.771	0.0007	0.00014	0.20014	7343.9	4.997	**40**
45	3657.258	0.0003	0.00005	0.20005	18281.3	4.999	**45**
50	9100.427	0.0001	0.00002	0.20002	45497.1	4.999	**50**
∞				0.20000		5.000	**∞**

About the Author

Robert L. Hershey is a Vice President of SMC Management Technology, a unit of the Science Management Corporation, an international management consulting firm, in Washington, D.C. He has held engineering and management positions at Booz, Allen and Hamilton; Bolt Beranek and Newman; Weston Instruments; Bell Telephone Laboratories; and Eastman Kodak. He is the author of more than 30 technical papers and reports in the fields of management, energy, environment, and operations research. Most of his recent work has involved research and development program planning for government and industry clients.

Dr. Hershey holds a Ph.D. degree in engineering from Catholic University of America (1973), an M.S. degree in mechanical engineering from the Massachusetts Institute of Technology (1964), and a B.S. degree *summa cum laude* in mechanical engineering from Tufts University (1963). He has been President of the D.C. Council of Engineering and Architectural Societies, President of the D.C. Society of Professional Engineers, President of the MIT Club of Washington, Chairman of the Washington Section of the American Society of Mechanical Engineers, and Treasurer of the Washington Chapter of the Acoustical Society of America. He is a member of various professional societies and organizations, including Tau Beta Pi (engineering), Sigma Xi (science), Omicron Delta Epsilon (economics), and Mensa.

Designed by Joan Lewellyn.
Cover design and page layout by Spectra Media.
Composed in Phototype Bookman by
The Bookmakers, Palo Alto, California.
Printed offset by the
George Banta Company, Harrisonburg, Virginia
on fifty-five pound Whitman
in an edition of 7,500 copies.